CLIMATE

ECO-SOCIALISM

JEREMY NIEBOER

www.brugesgroup.com

ABOUT THE AUTHOR

JEREMY NIEBOER was educated at Harrow School and Oriel College Oxford. After a period practising as a barrister in Kings Bench Walk he was admitted as a solicitor becoming a partner in two City law firms. He specialised in corporate work including mergers and acquisitions, capital market public offerings, private equity transactions and commercial law. He still acts for a few long standing clients.

His first encounter with any challenge to the accepted doctrine of 'global warming' came through his contact with Christopher Booker whom Jeremy first met when acting as lead solicitor on the application by Lord Rees-Mogg to restrain ratification of the Maastricht Treaty. Christopher himself published his essential work "The Real Global Warming Disaster" in 2009. Just at the time of its publication there was public meeting in Church House addressed by Professor Plimer in which he succinctly set out the fundamental scientific flaws of the theory of alleged CO_2 driven global warming. It was this that set Jeremy on a path of enquiry and research.

He has been a lead speaker at public meetings and debates on the want of any tenable scientific basis for the demonisation of CO_2 and the falsehood of global warming theory. He published his first booklet on climate alarmism with the Bruges Group in 2010 "A Lesson in Democracy". His book "CLIMATE All is Well All will be Well" was published in October 2021 and became 1st Best Seller on Amazon books in Ecology and the Environment. It was described by William Happer Professor Emeritus of Physics Princeton University USA as "*excellent*" and by Michel Van Biezen Professor of Physics at Loyola Mount University USA as "*impressive*".

In June 2022 his much praised second book "CLIMATE CO_2 Nature's Gift" was published by the Bruges Group as the second in a trilogy on the false ideology of climate change. It has been described as an "*Excellent contribution to the important debates about climate change and energy policy*".

"CLIMATE Eco-Socialism" is the final book of the trilogy.

FOREWORD

This is the third and final CLIMATE book. *All is Well All will be Well* explained why we have nothing to fear from burning fossil fuels and how the Sun, Earth's orbits and great ocean shifts govern temperature. *CO_2 Nature's Gift* showed that though CO_2 is our most beneficial gas providing all our oxygen and food it is at its lowest level for 99% of the time since life emerged. If CO2 doubled crop yields would rise by over 50% without any noticeable effect on warming.

But there remains an indelible unease that humanity has in some way disturbed the balance of Nature. Explaining that CO_2 cannot cause global warming does not dislodge the psychological effect of the 30 years of repeated predictions of disaster awaiting us.

The global warming dogma is uttered in the name of 'Science'. To challenge it seems to require knowledge of complexities beyond us. We are reluctant to dispute the need for extreme measures demanded by a form of eco-socialism. But there is a further and much deeper reason why it is not challenged. It is the phenomenon of the Big Lie.

"The broad masses of a nation are always more easily corrupted in the deeper strata of their emotional nature than consciously or voluntarily; thus they more readily fall victims to the big lie than the small lie, since they themselves often tell small lies in little matters, but it would never come into their heads to fabricate colossal untruths and they would not believe others could have the impudence to distort the truth so infamously. Even though the facts which prove this to be so may be brought clearly to their minds, they will still doubt and waver and will continue to think there may be some other explanation[1]."

The founder of global warming ideology himself knew that at two thirds current levels CO_2 has already absorbed all available radiated heat so that adding more has no noticeable effect - even if doubled it would result in warming of just 0.8^0C. Yet he colluded[2] with the official custodian of global temperature records in suppressing this basic principle of spectroscopy[3].

The intent behind this colossal deception may be inferred from its effects. The elimination of fossil fuel combustion which eco-socialism requires will lead to the collapse of the open market economies of those States that seek to enforce it.

This final CLIMATE volume raises four questions forced upon us by the insanity of Net Zero.

How have the liberal democracies of the World allowed the collapse of industrialised civilisation[4] to become even a possibility?

How is it that the dismal and fallacious conception of a fearful and imperilled humanity has been allowed to paralyse reason?

How has a policy rooted in a colossal deception come to be dictated by an unaccountable bureaucracy and basic negating principles of spectroscopy distorted and suppressed?

What States will benefit from such a disaster? Are they not China and Russia whose scientists and governments give no credence to the global warming deception[5]?

Will our way of life be preserved if we allow these things or do not even seek answers?

---

[1] Adolf Hitler Mein Kampf 1925.

[2] Stephen Schneider research paper 1971 *"Science"*. He worked with James Hansen director Goddard Institute. See below at Part 1 pp 6 – 8.

[3] Study of absorption and emission of radiation by matter involving splitting electromagnetic radiation into its constituent wavelengths.

[4] *Isn't the only hope for the planet that the industrialised civilisations collapse? Isn't it our responsibility to bring that about?"* Maurice Strong Promoting founder of the IPCC, UNEP and of Climate Change Conferences of the Partie. Interview 1992 Gibson, Donald. Environmentalism: ideology and power. p 95.

[5] See below p 63 footnotes 228 and 229.

ABSTRACT

Increasing CO_2 in the atmosphere cannot cause global warming[6]. Even at pre-industrial levels CO_2 has absorbed all available surface radiation– it is 'saturated'. For 50 years it has been known that adding further CO_2 has negligible effect. Evidence of this fundamental principle has been distorted and suppressed. It is the greatest deception in history.

No distinguished professor having expertise and experience in infrared spectroscopy[7] at a leading scientific institution who would now stake his career on the existence of conclusive evidence of global warming induced by rising CO_2 due to human combustion of fossil fuels.

Global warming due to human CO_2 emissions is the dogma of an unaccountable bureaucracy – the IPCC – laid down as incontrovertibly true. It has no scientific validity. It depends entirely on predetermined modelling and falsified records. It is contradicted in all respects by spectroscopy and actual empirical observation. It is maintained by deceit, fear and guilt.

Mankind's discovery of Nature's secrets heralded the modern age. In less than 50 years from 1800 to 1848 the English industrial revolution brought into existence the open market economy which fostered the means of credit and exchange that sustained it, the liberal democracy that shaped its form and the conditions of free trade that ensured its flowering, all at no cost to the global environment.

Developing nations owe to their open market economies a wellbeing that their inhabitants even in 1950 could not imagine. It has not been at the cost of injury to Nature. Population growth has accelerated it. Surplus food provides for those living in such economies and for most who do not. Energy stores are abundant. There were no cancer inducing pesticides. There was no acid rain. There will be no nuclear winter. There was no ozone hole. There will be no ice age. There is no global warming and there will be none.

The dominion of eco-socialist global warming dogma over science has subverted sincere human concerns induced by the shock of the profound transformation of human existence. It is a cult belief founded on dogma not evidence. Eco-socialism seeks to replace one nation democracy with a form of global governance, to subvert the open market economy and procure vast wealth re-distribution. It admits no contradiction or doubt. It is based on a colossal falsehood.

The root of a free society is the exchange of knowledge and ideas by interaction of individuals in spontaneous, unpredicted responses. Only with freedom of knowledge and industrial opportunity - only since everything could be tried and somebody found to back it at his own risk-- has science changed the face of the world. Democracy is the obstacle to the loss of freedom which eco-socialism necessarily requires.

On every measure of wellbeing -- life expectancy: elimination of disease: abundance of food: defeat of poverty: safety: security: tolerance: freedom: prosperity and human development - liberal democracy and open markets have within just four generations so transformed human existence as to inspire wonder and gratitude to a merciful and generous Providence.

---

[6] In this book the term "global warming" means heating of the surface atmosphere of the Earth by reason of man made emissions of CO_2 to such an extent as to imperil the very existence of the human race.

[7] Study of absorption and emission of radiation by matter by splitting electromagnetic radiation into its constituent wavelengths.

CONTENTS

PART 1	**THE BIG LIE**..	1
I	Radiation saturation of CO_2..	1
II	Radiation absorption..	3
III	Schneider: Fraud and Fabrication..	6
Summary	....	8
PART 2	**DOMINION OF DECEIT FEAR & GUILT**..	9
I	Deceit..	9
II	Fear..	17
III	Guilt..	20
Summary	....	22
PART 3	**THE COMING OF THE INDUSTRIAL AGE**..	23
I	The Great Transformation..	23
II	Origins and Causes..	27
III	Open Market Economy..	29
IV	Middle Class Democracy..	31
V	Free Trade..	33
Summary	....	36
PART 4	**MAN & NATURE 1792 – 1988**..	37
I	The Shock of Change..	37
II	Population and Food..	40
III	Depletion of natural resources..	44
IV	Disturbance of the balance of nature	48
Summary	....	53
PART 5	**EMERGENCE OF ECO-SOCIALISM 1988 – 2022**..	54
I	The Static and the Dynamic..	54
II	Rise of Eco- Socialism anti-Capitalism..	57
III	Maurice Strong. Conscription of the UN..	60
IV	Wealth re-distribution..	66
Summary	....	70
PART 6	**ECO-SOCIALISM & OPEN MARKET ECONOMY.**..	71
I	Opportunity and Insecurity..	71
II	Free Society and the Individual..	74
III	Functioning of the Open market economy..	76
IV	Eco-Socialism. Propaganda. Freedom..	79
V	Eco-Socialism v. Open market economy..	85
Summary	....	86
PART 7	**THE SUNLIT UPLANDS**..	87
I	Life expectancy..	87
II	Elimination of disease..	91
III	Food..	98
IV	Abolition of extreme poverty..	102
V	Safety and Security..	104
VI	Tolerance and Freedom..	105
VII	Human Development..	106
Summary	....	107

FIGURES

1. Image of the arrest of convection at tropopause 3
2. Infra red electromagnetic wavelengths 4
3. Planck curve and ‘greenhouse effect’ 4
4. Logarithmic decline of heating by ‘greenhouse effect’ 6
5. GISS 2007 Fabrication of US 20th century temperature record 11
6. IPCC 1990 Correct record of 1000 year global temperature record 11
7. IPCC 2001 Fabrication of 1000 year temperature record (M Mann) 12
8. GISS 2017 Fabrication of 20th century global temperature record 13
9. IPCC 2021 Fabrications of global 2000 year and 20th century temperature records 13
10. Greenland ice core record of 10,000 year temperature fluctuations 14
11. 570m years temperature and CO_2 fluctuations 15
12. GDP per capita UK 1270 – 2016 24
13. GDP UK, France and Germany 2000 years to 2022 25
14. UK Population 10,000 years to 2022 26
15. Coal production and imports 1750 – 2019 27
16. CO_2 comparative emissions 1850 – 2022 28
17. Tariff rates UK, USA and France 1850 – 2000 34
18. International trade 1500 – 1800 35
19. UK exports of goods re GDP 1827 – 2014 35
20. Globalisation of trade 1500 – 2017 36
21. Malthus curves of population and food 40
22. Global population 1800 – 2100 (UN prediction) 41
23. Famines 1860 – 2019 42
24. Food production re CO_2 emissions and population 42
25. North Sea coal deposits 46
26. Brandt line 59
27. IPCC 2007 Report modelled predictions of surface temperatures 67
28. Satellite record of global temperature 1979 –January 2023.... 68
29. Photographic image of cycles of solar maximum and minimum activity 83
30. Life expectancy global comparisons 1800 – 2019 87
31. Life expectancy global comparisons 1540 – 2019 88
32. Global Map of length of lives 89
33. Child mortality 1960 – 2020 90
34. Maternal mortality 1751 – 2020 90
35. UK maternal mortality 1847 – 2017 91
36. Small pox cases in London 1629 – 1902 92
37. Small pox – global reported cases 1920 – 2016 93
38. Poliomyelitis global reported cases 1980 – 2020 94
39. Deaths from measles in London 1629 –1902 97
40. Global reported cases of measles 1980 – 2021 97
41. Undernourishment in developing countries FAO data 1970 – 2015 100
42. Atmospheric CO_2 rise in atmospheric density 1750 – 2000 101
43. Long term wheat yields 1850 – 2014 in Europe 101
44. Global GDP 1820 – 2018 102
45. Global per capita comparative GDP 1820 – 2018 102
46. Extreme poverty index 1990 – 2019 103
47. Long term homicide rates Western Europe 1300 – 2016 104
48. Duration of wars between major powers 1500 – 2015 104
49. Democracies re other forms of government 1800 – 2020 105
50. Gender equality index 1950 – 2000 105
51. Tolerance of homosexuality 1791 – 2019 106
52. Index of Human Development 1870 – 2015 106
53. Literacy rate 1475 – 2015 107

SUMMARY OF CONTENTS

Part 1 Explains 'saturation' of atmospheric CO_2 - how CO_2 at pre-industrial levels of concentration has already absorbed all available radiation from the Earth's surface. Adding more CO_2 has negligible effect even at multiples of concentration. Despite being known for over 50 years that the physics of saturation precludes CO_2 causing global warming the extent of its impact has been distorted or concealed by the IPCC.

Part 2 Explains the propagation of global warming dogma by use of deceit and inducing of fear and guilt. The IPCC[8] and GISS [9] publish fraudulent graphs of temperature records of the past 1000 years and 20th century in order to maintain intimidation. Describes how fear and guilt induced by climate propaganda is seriously disturbing millions particularly children.

Part 3 Describes the emergence of the modern world with the coming of the industrial age in England and its shocks. Reviews the emergence and rise of the open market economy, banking institutions and investment exchanges. Traces the inception of middle class liberal democracy and its impact on the industrial economy. Reviews the effects of adoption of free trade.

Part 4 Analyses 19th century concerns as to perceived impacts of the industrial revolution as to population, food and energy resources. Discusses 'scarcity' and Thomas Sowell on the economics of resources. Reviews the alarms of 1947 – 1987 as to human intervention in Nature. Explains how none were justified by evidence of empirical science. Reviews the notion of "sustainability" and emerging eco-socialism.

Part 5 Describes how in just 4 years from 1988 – 1992 post-war environmentalism was subverted and replaced by eco-socialism and its single global warming dogma. Describes the work of Maurice Strong, the origins and creation of the IPCC and the extension of eco-socialism. Examines present day eco-socialism, its essential features, its principal advocates, its propaganda and networks.

Part 6 Analyses the open market economy in relation to eco-socialism. Reviews the works of Friedrich Hayek and Thomas Sowell as applied to modern economies and liberal democracy. Describes the role of democracy in open market economies. Examines how propaganda has secured submission to an eco-socialist global warming dogma.

Part 7 Shows how liberal democracy and open markets have transformed human existence on every measure of wellbeing --life expectancy: elimination of disease: food and famine: defeat of poverty: safety: security: tolerance: freedom: prosperity and human development.

Concludes that we live in a golden age and upon broad and sunlit uplands.

---

[8] Intergovernmental Panel on Climate Change
[9] Goddard Institute of Space Sciences (part of NASA) one of two official custodians of surface thermometer readings of temperature.

PART 1.
THE BIG LIE

"We want one single grand lie which will be believed by everybody including the rulers"
Plato's Republic Book 3 414b-c[1]

I. CO_2 'Saturation'

Climate Change Act

On 9th June 2008 the Climate Change bill was approved in the House of Commons of the United Kingdom by the votes of 463 MPs - 92.6% of those attending. Just 5 voted against it[2]. It became law on 26 November 2008. It provided that CO_2 emissions in the United Kingdom had to be cut by 80% of the 1990 level by 2050. In 2019 this was increased to 100% of the 1990 level. The cost of compliance estimated by the National Grid is £3,000,000,000,000 (£3 trillion)[3].

It had no justification whatsoever in science.

The Schneider 'Saturation' research paper

On 9 July 1971 there appeared in *'Science'*[4], the journal of the American Association for the Advancement of Science, an article by S. I Rasool Chief Scientist for Global Change at NASA and Stephen Schneider Professor of Biology and Global Change at Stanford University, California.

The article concerned the effect, if any, of rise in atmospheric CO_2 on global temperature.

> *We report here on the first results of a calculation in which separate estimates were made of the effects on global temperature of large increases in the amount of CO_2 and dust in the atmosphere.* ***It is found that even an increase by a factor of 8 in thc amount of CO_2, which is highly unlikely in the next several thousand years, will produce an increase in the surface temperature of less than 2^0K[5].***
>
> ***From our calculation, a doubling of CO_2 produces a tropospheric temperature change of 0.8^0K[6].*** *However, as more CO_2 is added to the atmosphere, the rate of temperature increase is proportionally less and less, and the increase eventually levels off. Even for an increase in CO_2 by a factor of 10 the temperature increase does not exceed 2.5°K.* ***Therefore the runaway greenhouse effect does not occur because the 15 micron CO_2 band, which is the main source of absorption, "saturates," and the addition of more CO_2 does not substantially increase the infrared opacity of the atmosphere.***

---

[1] G.R.F. Ferrari Professor of Classics University of California comments that "The lie is grand or noble (*gennaios*) by virtue of its civic purpose" He explains that the Greek word is also used colloquially as meaning a massive lie as in 'grand larceny'."
[2] Christopher Chope, Philip Davies, Peter Lilley, Andrew Tyrie, and Ann Widdecombe.
[3] National Grid "*Future Energy Scenarios*" July 2020.
[4] S.I Rasool. S.H.Schneider. *'Atmospheric Carbon Dioxide and Aerosols. Effects of Large Increases on Global Climate'*. SCIENCE. Vol 173 9 July 1971 173 pp138 – 141. Goddard Institute for Space Studies New YorkAffiliate of NASA. Peer reviewed. Bold type added.
[5] Less than 2^0C.
[6] The same as 0.8^0C.

'Saturation' occurs when CO_2 has effectively absorbed virtually all of the infra-red surface radiation within the available wavelength bands of the radiation spectrum. More CO_2 has negligible effect.

CO_2 concentration in the atmosphere is now 417 parts per million (ppm). The saturation has already taken place even before atmospheric densities of CO_2 at the pre-industrial level of 280 ppm are reached– see Figure 4 page 6. Rasool and Schneider were simply recording what was well known to atmospheric physicists. Increases in density of atmospheric CO_2 do not thereby also increase the warming effect of infra-red radiation from the surface of the Earth since all available radiation has been absorbed. Doubling CO_2 will have negligible effect on temperature.

In 1900 Knut Ångström[7] asked his laboratory assistant Herr J. Koch to measure the passage of infrared radiation through a tube filled with carbon dioxide. Koch reported that the amount of radiation that got through the tube scarcely changed when he cut the quantity of gas back by a third. It took only a trace of the CO_2 gas to "saturate" the radiation. He had discovered that in the bands of the spectrum in which CO_2 absorbed radiation it was so effective that even at very low densities adding more gas would make no noticeable difference whatsoever[8]

It is asserted by environmental activists that more CO_2 in the atmosphere means more global warming[9]. It is the foundation of the global warming dogma. It is, however, utterly false. Saturation renders global warming impossible as a matter of atmospheric spectroscopy. Earth's temperature has never risen in line with CO_2 either over geological ages or in the recent past.

Solar radiation balance

The Earth maintains its temperature by balancing day time solar radiation[10] (ultra violet short wave) in part by reflecting radiation from its surface (infra-red long wave) both day and night. To maintain the balance solar energy is also transferred by evaporation and to a small extent by convection before being radiated to space.

The main atmospheric gases (oxygen and nitrogen) are 'transparent' to incoming short wave solar radiation and also to outgoing long wave re-radiation from the Earth's surface. They have molecular structures that preclude the electromagnetic oscillation that allows absorption of radiation. However, water vapour and carbon dioxide[11] are 'opaque' to certain very limited wavelengths of outgoing long wave thermal infrared energy radiated from the surface - they are capable of absorbing it within these precise wavelengths.

Only 48% of the Sun's short wave energy reaches the surface of the Earth[12]. An aggregate of 18% is radiated back as long wave (infra-red) radiation. 12% escapes direct to space. However the balance of 6% is absorbed by greenhouse gasses by electromagnetic molecular dipole interaction.

---

[7] Professor of Physics Upsalla University Sweden.

[8] Ångström K (1900).*Ueber die Bedeutung des Wasserdampfes und der Kohlensäure bei der Absorption der Erdatmosphäre.* Annalen der Physik. **3** (12): 720.

[9] **'global warming' means warming of the Earth's surface due to human CO2 emissions that constitutes a severe adverse threat to human welfare.**

[10] NASA Earth Observatory "Climate and Earth's Energy Budget" January 14 2009. 29% of solar energy is reflected back by bright particles. 23% is absorbed by the atmosphere. 25% is transferred by evaporation and: 5% by convection. 12% is radiated back direct to space through the 'atmospheric window' of 9 – 13 microns see Figure 2. 6% is absorbed by water vapour and other greenhouse gasses.

[11] Also methane (0.00018%) and nitrous oxide (0.0000385%) in minute densities.

[12] See footnote 10 above

This is the 'greenhouse effect'. It occurs within an altitude of just 200 metres. It keeps the Earth at an average of 15^0 C when it would otherwise be a snowball at minus 18^0 C.[13]

The energy cools as it rises by convection through the falling pressure and temperature of the troposphere to the junction with the stratosphere – the tropopause - where it is radiated to space. This radiation occurs at the tropopause since it is there that the temperature is becomes stable and the stratosphere itself has a higher temperature due to formation of ozone and its absorption of ultra violet radiation. The tropopause brings to an end the process of convection as shown in Figure 1 with the flattening of convected H_2O thus allowing cold air to descend to the lower atmosphere. H_2O molecules are almost entirely confined to the lower part of the atmosphere - the troposphere -and do not account for radiation. With rising temperature CO_2 becomes the predominant molecule for cooling the atmosphere since it alone radiates the convected energy allowing cold air to fall to the surface thus completing the convection cycle.

CO_2 thus both maintains a habitable surface temperature and also cools the lower atmosphere.

Figure 1

CO_2 has two key characteristics. It only absorbs infra-red energy within very limited infra-red wavelengths. Moreover, the addition of more CO_2 to the atmosphere has a logarithmic limiting effect. It declines 'asymptotically'[14] in ever decreasing amounts. The more it increases the less are its effects - as correctly noted by Rasool and Schneider.

II. Radiation absorption

Infra-red wave length bands

The thermal radiation from the surface of the Earth is absorbed by greenhouse gasses according to the infra-red wavelengths of the radiation. For CO_2 these are primarily in the bands 14 to 16 microns[15] the dominant band being 15 microns. The following chart (Figure 2) shows the absorption wavelengths for CO_2 (red) and for H_2O (blue). The left hand scale is percentage

---

[13] "Greenhouse" is a misleading term Short wave heat from the Sun is able to penetrate greenhouse glass but when heat is reflected back it is long wave and cannot penetrate the glass. The greenhouse effect absorbs and re-emits the 6% surface radiation cooling as it rises to the stratosphere.

[14] Pertaining to a limiting value for example of a dependent variable when the independent variable approaches zero or infinity.

[15] Micron = one thousandth of a millimetre.

absorption up to total saturation. CO_2 only absorbs what H_2O does not take up. Three of the CO_2 bands are taken by H_2O which has many more wavelength absorption bands than CO_2.

Moreover the CO_2 absorbing bands of 1.9 to 2.1 and of 4 to 4.6 are at very low levels in the spectrum of radiation (Figure 3). The effectiveness of the 15 micron wavelength band (667.5 cm frequency), with its bending mode of molecular oscillation, is due to its proximity to the peak of the spectrum of radiation (Figure 3). Thus CO_2 has an approximate 23% overall greenhouse effect despite being just 3.9% as a percentage of all greenhouse gasses with H_2O comprising 95%.

Figure 2

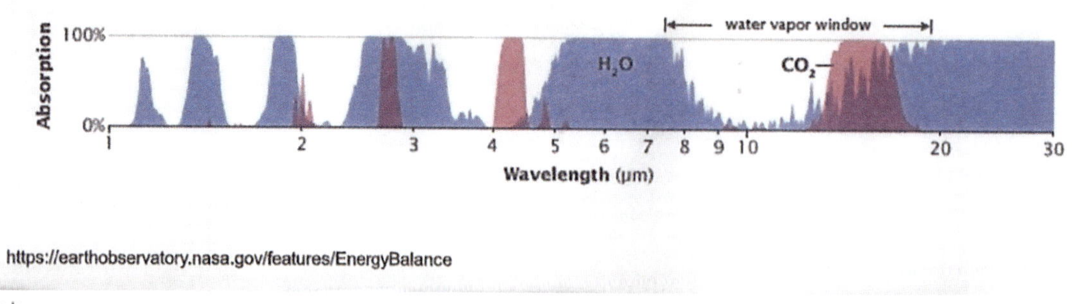

Climate sensitivity

The spectrum of the infra-red radiation emitted from the Earth's surface was defined by Max Planck. He found that radiation was emitted only in quanta. The spectrum is shown as the blue curve in the following graph.[16] The impact of emissions of radiation as attenuated by greenhouse gasses was defined by Karl Schwarzschild as the jagged black line within the Planck spectrum.

Figure 3

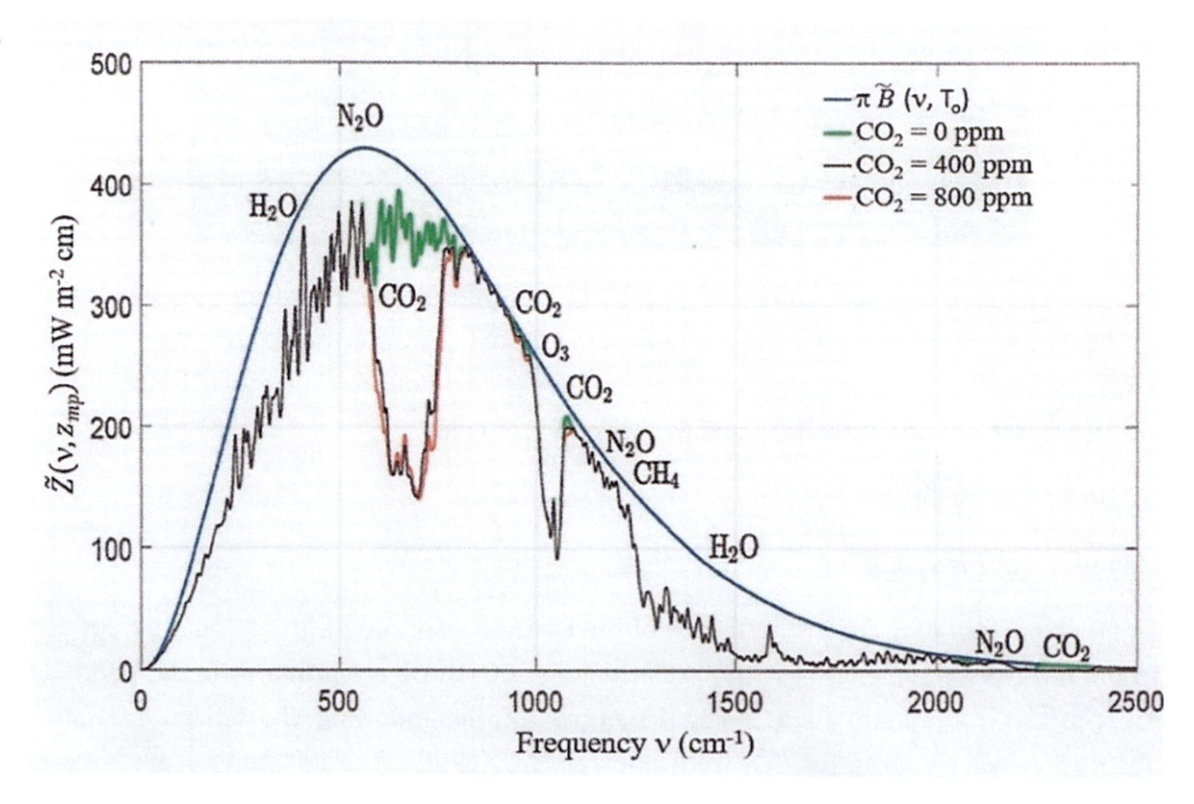

---

[16] Planck Curve named after Max Planck (1853 – 1947) who first discovered that adiated energy was emitted in quanta according to the Planck . constant governing the relationship between the energy of a photon and its frequency.

The horizontal scale is the frequency of thermal radiation. The vertical scale is the thermal power going to space at an assumed 15^0 C surface temperature (288K). If there were no greenhouse gasses the radiation into space would be all that is comprised within the blue curve.

The actual infra-red radiation from Earth to space is described by Schwarzschild's jagged black line in the above graph. The key element is the red line. This is what the Earth would radiate into space if CO_2 were to double its current atmospheric concentration.

The green line shows the extent of radiation to space in the assumed absence of all CO_2. The indented gap (bridged by the green line) is caused by CO_2 absorbing radiation that would otherwise go directly into space. If CO_2 density is doubled it does not double the gap – it just creates a tiny sliver between the black curve (base density) and the red curve (double density).

The Happer - Wijngaarden research papers

In June 2020 and March 2021 comprehensive research papers were published on the greenhouse gas effect by distinguished professors of physics William Wijngaarden of York University Toronto and William Happer of Princeton University USA. These set out fastidious analyses of the effect upon thermal infra-red radiation from the surface of the Earth of a doubling of CO_2 atmospheric density in a clear sky as well as of O_3, NO_3, N_2O and CH_4. The 2021 paper also covered CF_4 and SF_6[17] The authors used spectral lines and transition frequencies of over 1.3 million (2020 paper) and 1.5 million (2021 paper) rovibrational[18] lines from the most recent databases[19] in order to calculate the per-molecule 'forcing'[20] of the greenhouse gas molecules. The calculated findings coincided precisely with observational satellite measurements.

They found that the effect of doubling CO_2 would be an increase of 3WattsM^2–approximately 0.85^0C[21]. In other words, just as Rasool and Schneider had reported 50 years earlier. Moreover the Happer/Wijngaarden papers assumed a clear sky. When absorption and reflection by clouds are taken into account it results in an even lower level of average increase in surface heat.

Saturation - Logarithmic decline of heating effect

The absorption capacity of CO_2 declines logarithmically with increasing density - again as correctly reported by Schneider. The first 20 ppm of CO_2 operating as an atmospheric greenhouse gas has the most potent effect on temperature. Thereafter it has a declining capacity to absorb. At 280 ppm – the pre-industrial level - CO_2 has absorbed virtually all available infra-red radiation' The following diagram tracks this decline and each of the significant CO_2 concentration levels.

---

[17] Sulphur Hexafluoride SF_6 and Carbon Tetrafluoride CF_4..

[18] CO_2 and other trace gas molecules absorb electromagnetic radiation by its interaction with their electric dipole moments. These are the measure of the separation of positive and negative electrical charges or overall polarity. The process involves vibrational and rotational modes of oscillation. Rotational–vibrational spectroscopy is concerned with infrared and Raman spectra of molecules in the gas phase. Transitions involving changes in both vibrational and rotational states are rovibrational' (or ro-vibrational) transitions. When these emit or absorb photons (electromagnetic radiation), the frequency is proportional to the difference in energy levels and can be detected by certain kinds of spectroscopy.

[19] HITRAN (an acronym for High Resolution Transmission) molecular spectroscopic database used to simulate and analyse the transmission and emission of light in gaseous media, with an emphasis on planetary atmospheres. VAMDC (Virtual Atomic and Molecular Data Centre) atomic and molecular (A&M) data compiled within a set of AM databases.

[20] A climate 'forcing' is any influence on climate that originates from outside the climate system itself. The climate system includes the oceans land surface, cryosphere, biosphere, and atmosphere. Examples of external forcings include surface reflectivity (albedo); human Induced changes in greenhouse gasses; atmospheric aerosols (volcanic sulphates, industrial output).

[21] Professor Wijngaarden *"Greenhouse gasses contribution to 21st C warming"* 2022 Tom Nelson YouTube presentation

Figure 4

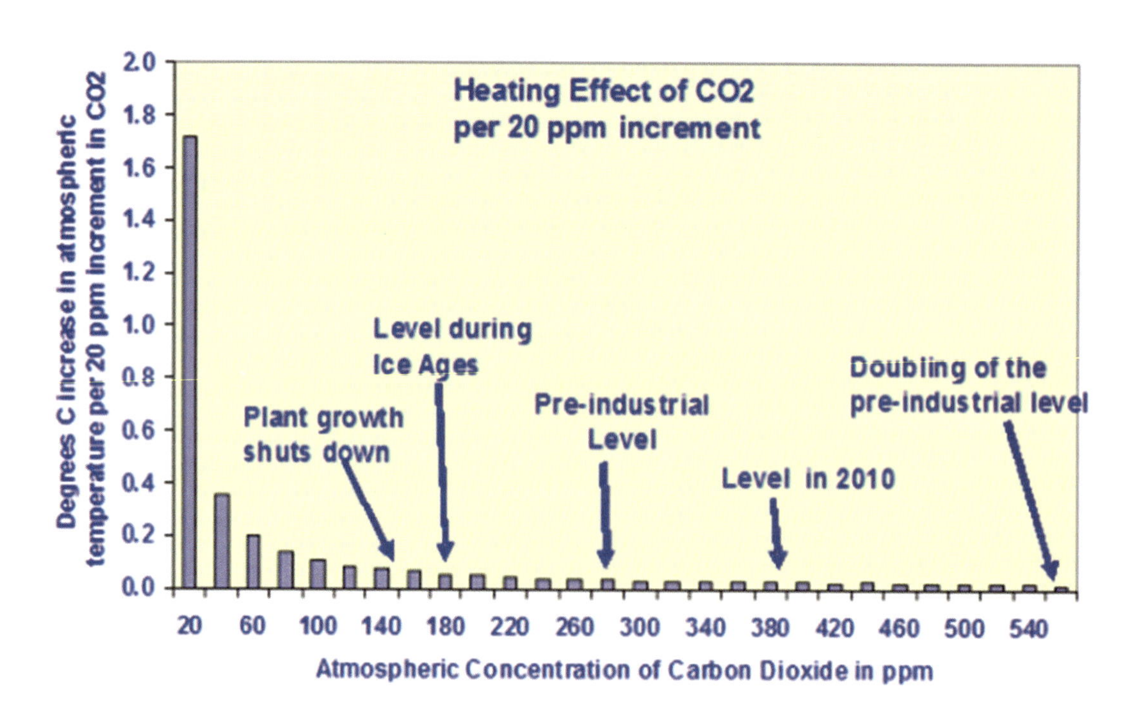

An analogy of this phenomenon would be the use of sunglasses. The first pair blocks say 60% of sunlight. Two pairs block 24%. The next pair and it is 14.4%- the next it is 8.6% and so on[22].

The IPCC have attempted to suggest that the minute warming contribution of CO_2[23] results in increased water vapour due to evaporation thus creating indirect leverage of the greenhouse effect of H_2O. This is not even arguable. There is no observational evidence whatever to sustain it. There has never been runaway global warming even when CO_2 was 10 times present levels in past epochs. It conflicts with Le Chatelier's principle[24]. Moreover water vapour (humidity) has not increased. It has decreased over the past 75 years[25] at all levels of the troposphere.

III. Schneider: Fraud and Fabrication

"Fame is one thing. Integrity is everything"

Stephen Schneider became the leading proselyte of the new global warming dogma. It was Schneider *"more than any other person created and drove the biggest deception in history"*[26].

Schneider's Global Cooling

In 1976, five years after his article in 'Science', Schneider reported, in a major work '*The Genesis Strategy*'[27], that the last 200 years had been unusually warm compared to the last 1,000 years but that 'this warm period' had now passed[28]. He asserted that the evidence pointed to the imminence of an ice age with disastrous consequences for food yields and famine in undeveloped countries. The front cover of the journal 'Science News' depicted the catastrophic effects.

[22] See also pp 373 – 375 Prof Ian Plimer *Heaven and Earth* 2009 Quartet Books published 12 years before the Happer-Wijngaarden paper.

[23] CO_2 is responsible for 23% of the greenhouse effect which absorbs 6% of solar heat reflected from the surface – a net 1.4%.

[24] Le Châtelier's principle states that if a dynamic equilibrium is disturbed by changing the conditions, the position of equilibrium shifts to counteract the change to re-establish an equilibrium.

[25] *Atmospheric relative humidity at various atmospheric pressures*. Climate 4U Professor Ole Humlum. Data source: Earth System Research Laboratory (NOAA). Diagram update: 11 March 2022.

[26] Watts up with That. April 11 2019 Dr Tim Ball.

[27] Schneider S. *The Genesis Strategy* February 1, 1976. Springer ISBN10: 03063090.

[28] Deborah Shapley article in New York Times July 18 1976.

'Newsweek' was among many publications to report on the scientific evidence. Schneider's prediction was reported as reflecting the 'climatological consensus'[29].

It goes further than that. Schneider gave his full endorsement of a 1976 book by US science writer Lowell Ponte *'The Cooling: Has the Next Ice Age Already Begun?"*. Schneider claimed that *"It is a cold fact [that] the Global Cooling presents humankind with the most important social, political and adaptive challenge we have had to deal with for ten thousand years".*

13 years later, in 1989, the year after James Hansen launched man-made global warming dogma into the political arena with his testimony to Congress,[30] Schneider published his *"Global Warming: Are We Entering the Greenhouse Century?"* He envisaged that the man-made CO_2 induced 'greenhouse effect' would give rise to a rise in global temperature of 5^0C and that this could happen as soon as 2050 unless drastic action was taken by industrialised countries[31].

Hansen joined Goddard Institute of Space Sciences (GISS) in 1967 later becoming its director. Schneider's 1971 article was published by GISS where he was a postdoctoral fellow. It confirmed that radiation saturation of CO_2 made it impossible for it to cause global warming. It was based upon computer models of Hansen himself.[32] Yet in 1988 Hansen and Schneider launched the man-made CO_2 dangerous global warming dogma knowing that saturation utterly excluded it.

Abuse of Scientific Method

Schneider was at ease with the practice of deceit in order to persuade. He was indifferent to whether he was the champion of Global Cooling or of Global Warming. He worked with Hansen on each of these falsehoods. He used his professional position and status to concoct an intellectual justification for extorting certainty from ambiguity.

Schneider spoke freely and in public about this. At a conference in 1996, of which he was co - chairman, he justified the creation of certainty from uncertainty as being absolutely necessary to provide policy prescriptions for global environmental change[33]. On 4 February 2010, shortly before his death, Schneider gave a lecture at Stanford[34] in which he dismissed the fundamental principle of the scientific method in determining the validity of a theory – that of falsification[35]. He proposed a new form of "science". This was his "*system science*". As he asserted:-

'Climate science is not like test tube science... we do not falsify by single experiments. We falsify on the basis of accumulated numbers of papers and numbers of bits of information".

Schneider understood completely the phenomenon of saturation of surface radiation by CO_2. He knew that increasing atmospheric CO_2 even by as much as a factor of 8 would have negligible

---

[29] Schneider Dr Tim Ball *"The Person who set the stage for Entire Deception of Human-Caused Global Warming"*. Watts up with That April 11 2019 quoting article in New York Times July 18, 1976.

[30] See Part 5 at p61. Hansen gave evidence to Congress on 23 June 1988 that he was 99% certain of man made dangerous global warming.

[31] *Global Warming: Are We Entering the Greenhouse Century*?" 1989 VintageBooks Random House pp 77 and 243.

[32] Washington Post 9 July 1971 'US Scientist predicts new Ice Age".

[33] Aspen Institute 1996 'Characterising and Communicating Scientific Uncertainty' "*How can science be most useful to society when evidence is incomplete or ambiguous, subjective judgment of experts vary and policymaker seek guidance and justification for courses of action. How can individual subjective judgments be aggregated into group positions"*

[34] Schneider S. Professor Environmental Studies and of Biology at Stanford University California.

[35] The scientific method established by Karl Popper. However many confirming instances there are for a theory, it only takes one counter observation to falsify it. Science progresses when a theory is shown to be wrong and a new theory is introduced which better explains the phenomena. Thus the scientist should try to disprove the theory rather than attempt continually to prove it.

consequences. He had succinctly described the phenomenon in his article in 'Science' of 1971[36]. It had been known since 1900 when discovered by Knut Johan Ångström. It was reiterated in 1981 by Professor Hoyle in his best selling book 'Ice'[37].

> *The efficiency of the carbon dioxide trap is insensitive to the amount of carbon dioxide in the atmosphere: increasing the amount five-fold would scarcely change the trap, in spite of the stories that are currently being circulated by environmentalists.*

In the 357 pages of Schneider's book on global warming there is no mention at all of saturation. He knew that the so-called 'system science' he was proposing was utterly falsified by the laws of spectroscopy and physics. Never at any moment since the emergence of multi cellular life on Earth 570m years ago had there been runaway warming for the reason that saturation by CO_2 rendered it impossible.

Schneider knew that saturation was undeniable and conclusive. He debauched science by contorting the scientific method into a charade of opinion forming.

There is no climate crisis.[38]

Justification of deceit

Schneider described his mendacious and manipulative *'system science'* in an interview with "Discover" magazine in 1989.

"On the one hand we are ethically bound to the scientific method, in effect promising to tell the truth, the whole truth and nothing but. On the other hand we are not just scientists but human beings as well….and would like…to reduce the risk of potentially disastrous climate change. To do that we have to ..capture the public's imagination … we have to offer up scary scenarios, make simplified dramatic statements and make little mention of any doubts we might have. Each of us has to decide what the right balance is between being effective and being honest".[39]

It is clear which of these options was chosen by Schneider.

SUMMARY

Increasing CO_2 in the atmosphere cannot cause global warming[40]. Even at pre-industrial levels CO_2 has absorbed all available surface radiation– it is 'saturated'. For 50 years it has been known that adding further CO_2 has negligible effect. Evidence of this fundamental principle has been distorted and suppressed. It is the greatest deception in history.

No distinguished professor specialising in infrared spectroscopy at a leading scientific institution who would now stake his career on the existence of conclusive evidence of global warming induced by rising CO_2 due to human combustion of fossil fuels.

---

[36] Schneider S et al *'Atmospheric Carbon Dioxide and Aerosols: Effect of large increases on Global Climate'*1971Science 173 138 141.

[37] Hoyle F. *'Ice'* 1981 Hutchinson & Co 1981 at p123 Sir Frederick Hoyle held the Plumian chair of Astronomy and Experimental Philosophy - one of the major professorships in Astronomy at Cambridge University in England.

[38] The scare has been compared to a curfew imposed on children to protect them from ghosts reported to be escaping from cemeteries.

[39] Press interview with the magazine *"Discover"* 1989

[40] In this book the term "global warming" means heating of the surface atmosphere of the Earth by reason of man made emissions of CO_2 to such an extent as to imperil the very existence of the human race.

PART 2

DOMINION OF DECEIT FEAR & GUILT.

I. Deceit

Never in all of the 35 years in which it has been uttering its propaganda has the IPCC published a definitive research paper giving an account of evidence of spectroscopy or other analytical explanation as to the molecular absorption by CO_2 and other greenhouse gasses of infra-red electromagnetic radiation from the Earth's surface and the constraints acting on its extent of quanta, frequencies, saturation, mutual impact of competing gasses and otherwise.

In the many thousands of pages that comprise its 'reports' of climate change dogma all that can be found are preconceived conclusions from models, surface thermometer readings and proxies each of which have been cynically formulated, altered or simply selected to procure them.

In 2001 and again in 2021 in its "Summary for Policymaker" it issued graphs alleged to have been derived from proxy evidence that were simply fabricated. They were the centrepieces of its dogma of dangerous global warming due to human CO_2 emissions.

The reason for this appalling debasing of science is that the whole fabric of the dogma is itself grounded on a colossal and deliberate falsehood. It follows that it can only be defended by an army of lies. The entire crusade to eliminate the basis of the modern economies of the West is maintained by three means, each characteristic of the propaganda of a totalitarian state[41].

These are Deceit, Fear and Guilt.

The essential justification of propaganda is that laudable ends warrant dishonest means. It is the 'noble lie' of Plato's Republic[42]. It informs the conduct and utterances of Schneider and of the thousands of the fervent disciples who mourned his loss and recited his catechism.

The dogma of global warming gained its grip upon the world in just 4 years from the announcement of the 'science' on 23 June 1988[43] to the adoption by 154 nations of the world of the UN Framework Convention on Climate Change on 12 June 1992.

It had no more basis in science than any of the alarms and scares of the previous 30 years. These had included panic that pesticides caused runaway proliferation of cancer: destruction of forests by acid rain: extinction of life due to soot from nuclear explosions: a new Ice Age: chlorine compounds causing a hole in the ozone layer exposing humans to risk of fatal cancers[44].

Falsification

The critical feature of all but one of these scares was that they were capable of being falsified[45] by empirical evidence of actual, present and existing physical phenomena. Indeed each of them were ultimately falsified in this way. However it was otherwise with predictions of future

---

[41] See the full examination of this use of propaganda set out in Part 6

[42] *We want once single grand lie which will be believed by everybody- including the rulers"* Socrates in Plato's Republic Book 3 414b-c

[43] Testimony of James Hansen Director of Goddard Institute for Space Science to Congress

[44] See Part 4 pp 37 -48 for a detailed examination of each of these scares and how they were falsified.

[45] See footnote 34 p7 for the famous test of Professor Popper as to validity of scientific hypotheses. *The Logic of Scientific Discovery* (1934).

catastrophic temperature shifts. For these models alone could sustain the forecasts.

Schneider knew that by concealing contradictory evidence and resting dangerous global warming theory entirely on modelled predictions of the future it could not then be falsified. As he himself said *"The future is always subjective."*[46] This also meant that his predictions could not be verified.

By 1975 hysteria was mounting at the catastrophe of an imminent Ice Age which was deemed to account for the steep decline in global temperature over the previous 30 years. 1946/47 and 1962/3 had been the coldest winters since 1740. Yet all the variations of temperature that were used to justify apocalyptic predictions of a new Ice Age were in reality natural cyclical changes accounted for by the state of solar activity and massive, natural and regularly occurring oceanic shifts. Schneider's ice age theory could not be falsified by observation and testing. No scientific evidence existed to support the hypothesis. It was futurology.

This was the identical scenario when, only a matter of 13 years later, the global warming hysteria was conjured up by Schneider and Hansen. Having no evidence to sustain it the adherents of the cult of global warming were compelled from the outset to utter exaggerated forecasts of doom and then to use fabrication and deceit as a principal means of reinforcing the indoctrination.

Falsehood becomes legitimate

Nor was there any reticence about this shocking debasement of science. There have been many brazen assertions. A few of them require repetition in order to underscore their iniquity.

"The data doesn't matter. We're not basing our recommendations on the data. We're basing them on climate models"[47] Professor Folland, Head of the Hadley Centre's Climate Variability and Forecasting Group from 1990-2008.

"The models are convenient fictions that provide something very useful". David Frame Deputy Director of the Oxford University Smith School of Enterprise and the Environment.

"No matter if the science of global warming is all phony ... climate change provides the greatest opportunity to bring about justice and equality in the world." Christine Stewart, former Canadian Minister of the Environment.

"It doesn't matter what is true; it only matters what people believe to be true" Paul Watson the co-founder of Greenpeace.

Tampering with records: removal of 1940-1976 cooling

Fluctuations in the temperature of the Earth during interglacial periods[48]are governed by cyclical solar activity[49], Earth's orbits and by shifts of heat in the mid[50]and north[51] Pacific. Since 1979 satellites and balloon sondes have provided accurate and comprehensive daily temperature data but for earlier years data is derived from surface thermometer readings. These records are kept

---

[46] Darwell R *The Age of Global Warming* p338 2013 Quartet Books an excellent account of the history of the scam – also his *Green Tyranny* 2017

[47] http://wattsupwiththat.com/2014/08/07/hoodwinking-the-nation-on-climate-issues/#more-114211.

[48] Periods of 11,000 – 15,000 years between ice ages (glaciations) which themselves last 80,000 – 100,000 years.

[49]In cycles of 11 years the Sun's activity waxes and wanes as magnetic fields inside the Sun burst through periodically so creating magnetic polarities which result in solar flares, coronal mass ejections and other electromagnetic phenomena.

[50] The El Nino and La Nina shifts of warm water between the International Date Line and 120°W including off the coast of South America.

[51] The Pacific decadal oscillation is a robust, recurring ocean-atmosphere climate variability in the Pacific Ocean, north of 20°N.

by the Goddard Institute of Space Science (GISS) and the UK Met Office Hadley Centre Climate Research Unit (HADcrut). There have been repeated falsifications of temperature records by both GISS and HADcrut[52]. The following are just a few examples. The sharp temperature increase of the 1930s and the following severe cold from 1940 to 1976 invalidated IPCC claims that temperature increased in line with rise in CO_2. In 2007 the GISS official graph of surface temperature was therefore altered from the 1999 GISS record (left graph) showing a cooling trend since the 1930s to show a steepening overall warming trend (right graph).

Figure 5 **1999** **2007**

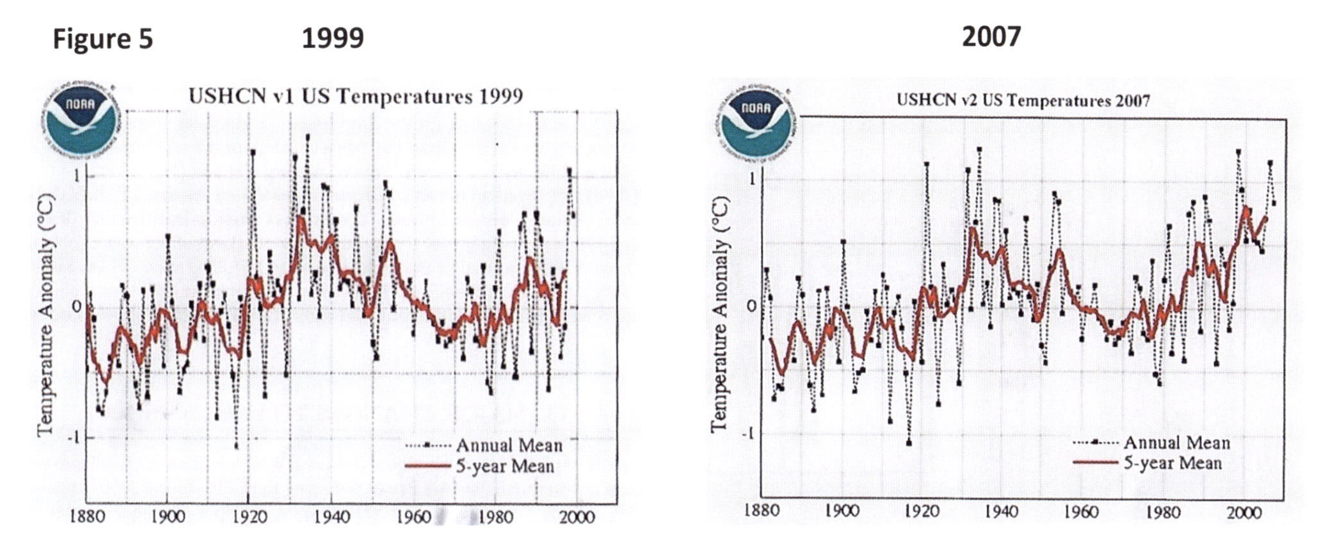

But the collapse in temperature of 1940 – 1976 remained. Thus e-mails appeared in 2009 from HadCRUT revealing cynical designs on the eradication of this established fact. One such included "*It would be good if we could get rid of the 1940s land blip*"[53]. It is unsurprising that other climatologists find that global surface records "*are nothing but a propaganda tool*"[54].

Removal of Medieval Warming and Little Ice Age. The first 'Hockey Stick' IPCC 2001

The IPCC's first report (1990) had included a correct graph of climate change over the last 1,000 years showing the Medieval Warming period and the severe Little Ice Age that followed it.

Figure 6

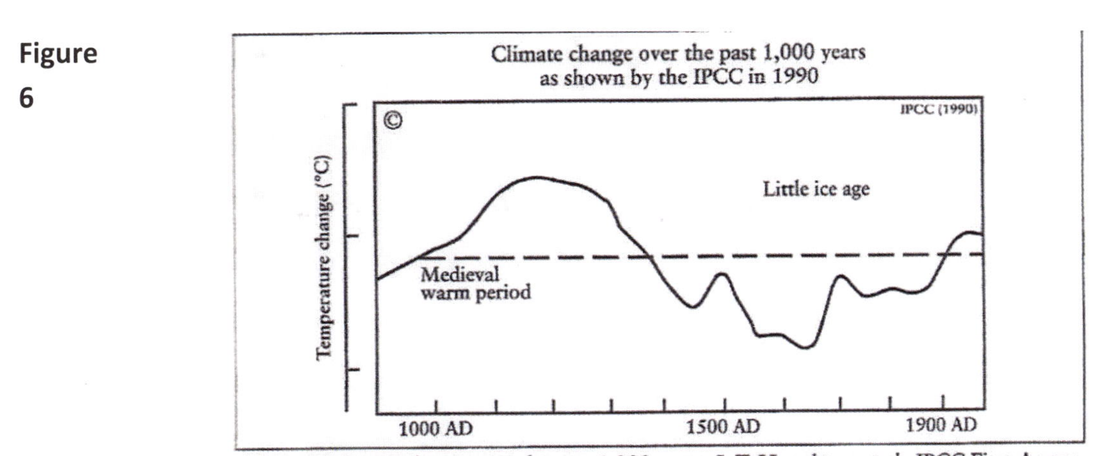

Figure 2: Climate change over the past 1,000 years: J. T. Houghton *et al.*, IPCC First Assessment Report, 1990.

---

[52] Booker C *The Real Global Warming Disaster* Continuum 2009. Dr Ball T. 2014 *The Corruption of Climate Science* Stairway Press.

[53] E-mail Tom Wigley of the University Corporation for Atmospheric Research wigley@ucar.edu to Phil Jones p.jones@uea.ac.uk. Subject: 1940s Date: Sun 27 Sep 2009 23:25:38 cc Ben Santer santerl@llln.gov.

[54] Dr Mototaka Nakamura *The Global Warming Hypothesis is an Unproven Hypothesis*" 2019.

Then in 1997 a survey[55] reported that 90% of US State Climatologists had found that *global temperature fluctuated naturally and in cycles over long periods of time"* when there were no human emissions whatsoever. This sabotaged the IPCC hypothesis. Professor Overpeck[56] therefore warned Professor Deming[57] that *"we have to get rid of the Medieval Warm Period"*.

Thus in 1998 there appeared in the journal 'Nature' a graph prepared by a PhD researcher, Michael Mann. It eliminated all the steep fluctuations over 1000 years as depicted in the IPCC 1990 report (Figure 6) and showed a sudden volcanic rise from about 1930 - the "hockey stick".

Figure 7

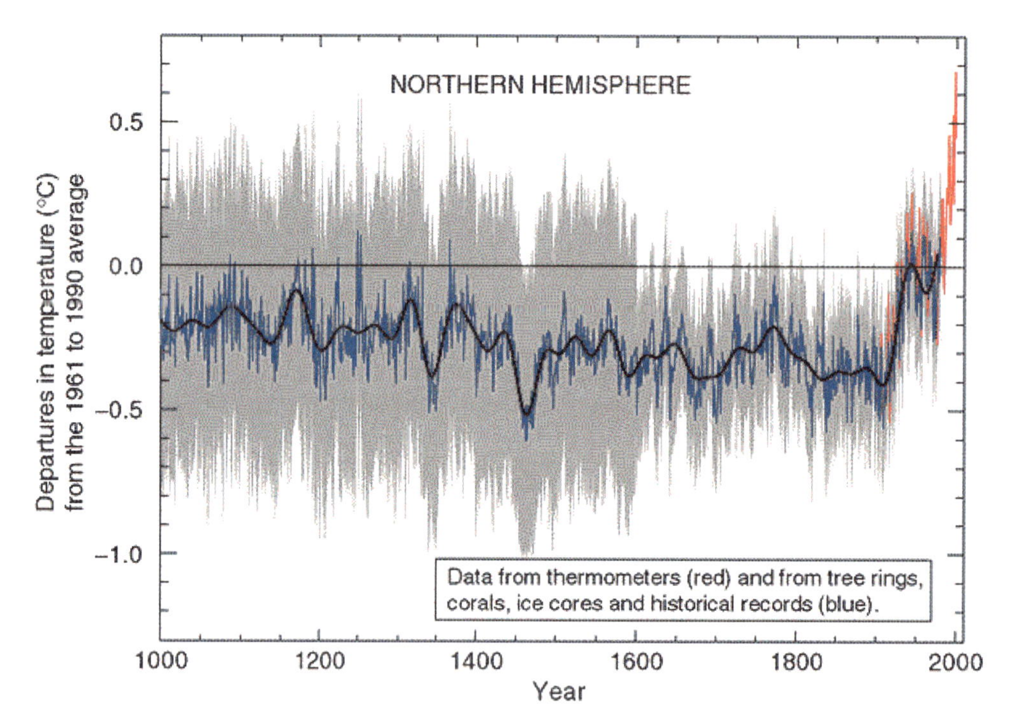

The hockey stick graph formed the first page of the key "Summary for Policymakers" of the 2001 IPCC 'report'. It was repeatedly relied on in the text of the 'report' and a vast depiction of it was showcased at the launch of the 'report'. One of its models now predicted temperature rise of no less than 5.8°C by 2100 - over 8 times the recorded increase for the entire 20th century (0.7°C). Mann simply eliminated the 1940 – 1976 "land blip", the Medieval Warming Period and the 500 years of the severe Little Ice Age 1310 – 1810.

Mann[58] had used an algorithm that gave 390 times more emphasis to Californian bristlecone pines which had anomalous 'hockey stick' patterns. The algorithm had been programmed to 'mine' for hockey stick shapes whatever data was introduced. Two Congressional inquiries as to Mann's graph resulted in its being utterly discredited.

---

[55] *"Survey of State experts casts doubt on link between human activity and global warming"* Citizens for a Sound Economy Press Release 1997.

[56] University of Arizona Climate System Scientist. Overpeck was one of a small group working closely with the IPCC. He was in Working Group 1 on the Summary for Policymakers. He was later Coordinating Lead Author for IPCC 4th Assessment (2007).

[57] A geoscientist at the University of Oklahoma.

[58] Mann held a position at Pennsylvania State University Earth System Science Center. Dr Timothy Ball was an environmental consultant and climatology professor at University of Winnipeg, Manitoba, with a doctorate in climatology from Queen Mary College University of London, He was Chief Science Advisor of the International Climate Science Coalition and Policy Adviser to The Heartland Institute. He commented that *"Michael Mann should be in the State Pen not Penn State"*.

Falsifying of 20th temperature record. GISS 2017 fabrications

Hansen as director of GISS (an affiliate of NASA) had in 1981 set out, correctly, the fluctuations of temperature of the period 1900 to 1981 in a simple but accurate graph (Figure 8).

This was inconsistent with the gradual ascent of temperature necessary to sustain the man made global warming theory that was fabricated 7 years later. Accordingly in 2017 GISS, under its new director, baldly reconstructed the record and reduced the warming of 1880 to 1940 from 0.8⁰C to 0.4⁰C with reduced the cooling from 1940 to 1975 by 0.3⁰C to align the trend with rise in CO_2.

Figure 8

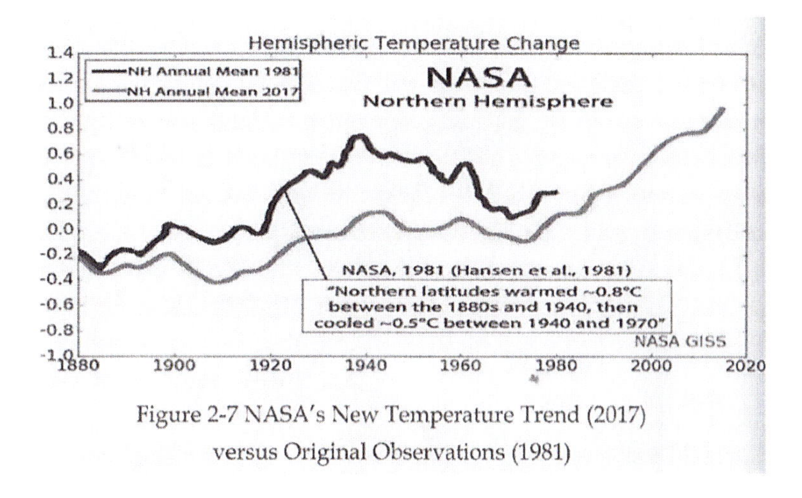

Figure 2-7 NASA's New Temperature Trend (2017) versus Original Observations (1981)

The 2021 IPCC Hockey Stick. Fraudulent graphs a) and b).

The IPCC again issued fabricated temperature records in its 2021 Summary for Policymakers.[59]

Figure 9

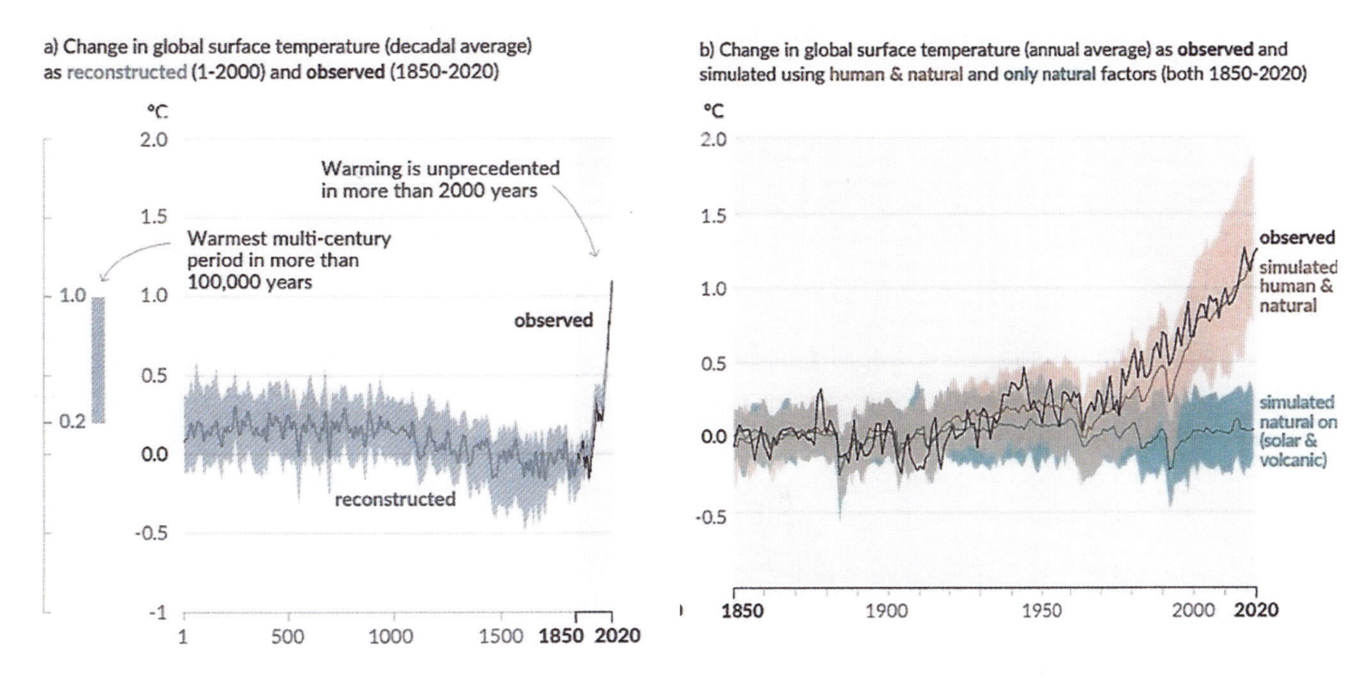

It sets out two graphs (above). Panel a) shows a very slight gradual decline of temperature over 2,000 years with a vertical rise since 1960. Panel b) shows minor variations from 1850 with the same

[59] See Steven MacIntyre Climate Audit August 2021 "*The IPCC AR6 Hockeystick*".

steep rise in the last 60 years. The Summary asserts that the warmest multi-century period in at least the last 100,000 years was around 6,500 years ago and that *"The last Interglacial, around 125,000 years ago, is the next most recent candidate for a period of higher temperature"*.

The IPCC findings depend on two serious fraudulent misrepresentations.

That temperature is now the highest of our 11,000 year interglacial period - only exceeded by a peak 6,500 years ago and then by the prior interglacial period 125,000 years ago.

That global temperature was in gradual decline from 1 AD until 1960 AD with only slight variations of 0.3°C but with a very steep rise since 1960 of 1.25°C.

Figure 10

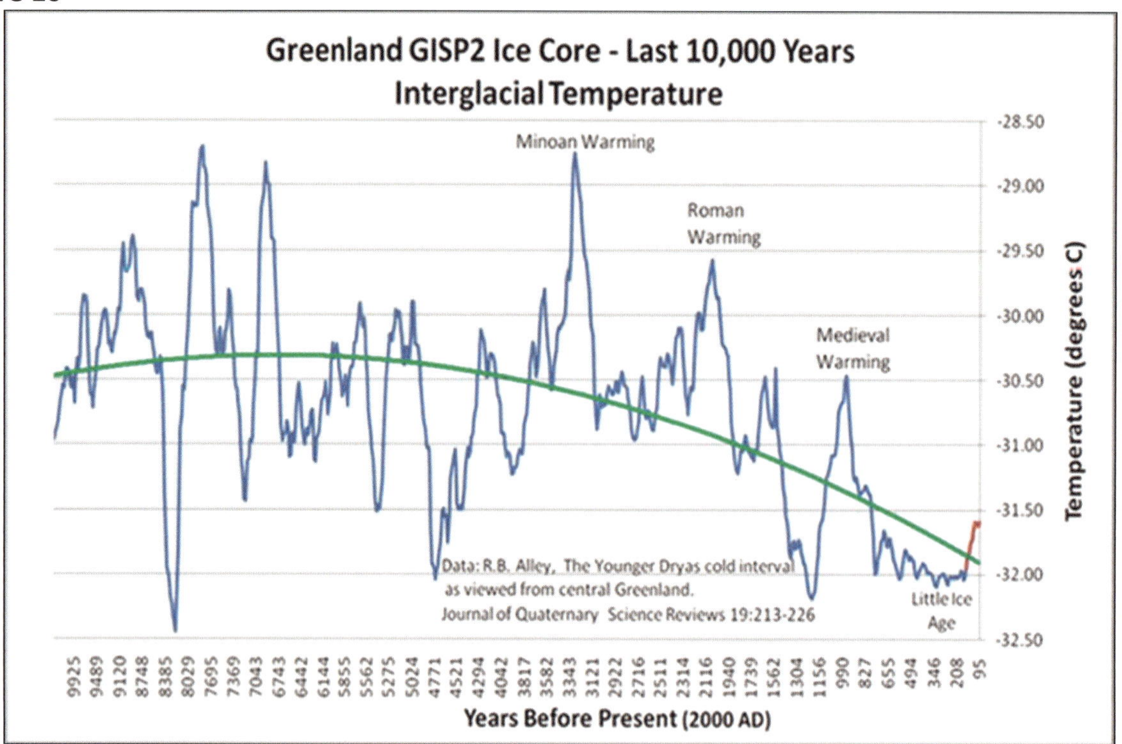

The above diagram depicts the variations on temperature the period since the last ice age. The maximum temperature of the last 10,000 years was 8,000 years ago. The statements that temperature is higher than for 6,500 years ago and then for 125,000 years are fraudulent.

In the past 11,000 years there have been very prolonged periods in which temperature was far higher than today (Figure 10). For the previous 100,000 years there was an Ice Age and accordingly temperature had fallen to far lower levels due to glaciation.

The IPCC left hand graph a) from 1 AD simply removes the fluctuations of the Roman Warming period, the Medieval Warming period and the Little Ice Age (Figure 6). The right hand graph b) from 1850 AD smothers the peak of the 1920s/1930s warming, not matched until the freak year of 1998. The severe cooling of the 30 years from 1945 has been cynically ignored.

The intention of these fraudulent truncations is to show 'Policymakers' that temperature had been in modest decline over thousands of years ending with a terrifying rise just 60 years ago. These are statements that the IPCC know to be false. They are made with the intention that they are to be relied on. They have been relied on to the grave detriment of whole societies and States.

The fabrication of the '97% of scientists' consensus

The IPCC depends not on the evidence of empirical observation and experiment but on manufactured models and alleged 'consensus.' A 2010 survey[60] by a student claimed that 97% of 'climate scientists' agreed with the IPCC dogma on global warming. The claim is re-cycled as evidence of a global scientific 'consensus'. In 2019 that claim was elevated to 100%. Thus there was not a single scientist with expertise in spectroscopy, electromagnetic radiation, orbital cycles and molecular properties of CO_2 who thought temperature variations were naturally caused.

Research proved[61] that the survey questions were asked of 10,257 "Earth scientists". But the survey authors decided that the disciplines of most respondents did not qualify them to answer, including physicists, geologists, astronomers and experts on solar activity[62]. The survey was then restricted to 3,146 'Earth scientists'. When this failed to secure the desired result the survey was limited to 79 in the 'Earth science' community. 77 agreed that human activity had contributed to warming. By dividing 77 by 79 the figure of 97% arrived. - 0.75% of the total claimed.

In testimony to US Congress a lead author of the IPCC[63] stated that "*The 97% is essentially pulled from thin air; it is not based on any credible search whatever*". It is a fabrication. Yet it is the consensus that has stupefied rational processes of enquiry.

Carbon dioxide a "Pollutant"

There is now a famine of CO_2.

Figure 11

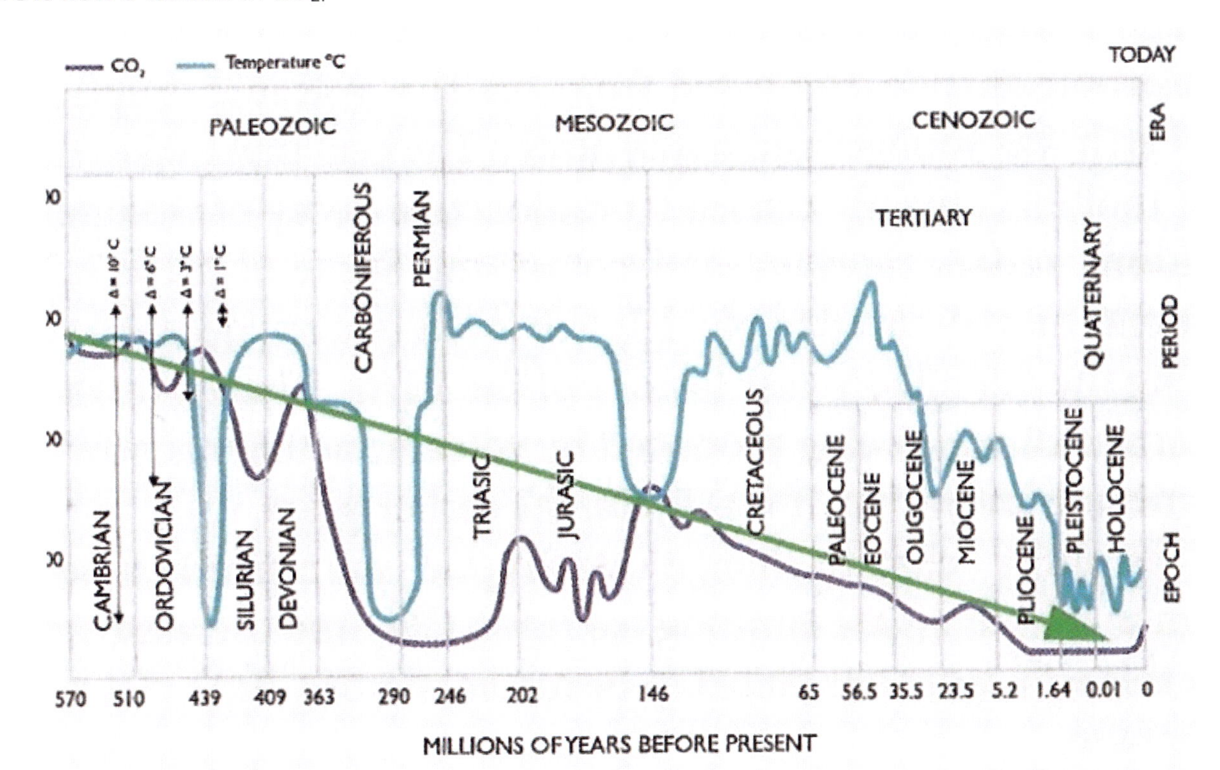

[60] See Booker C *Global Warming A Study in Group Think* p51 2017 Global Warming Policy Foundation. Enquiry revealed that the survey had been prepared by a Masters degree student at the University of Illinois.
[61] For a good analysis see Professor D Easterbrook "*Evidence Based Climate Science*" 2nd edition pp3,4. Elesevier Inc.
[62] Yet these are precisely the disciplines that most closely bear on issues of climate.
[63] Dr Richard Toll testimony to US Congress: Full Committee Hearing – Examining the IPCC Process 29th May 2014.

The graph above (Figure 11) is a record of the 570 million years since multicellular life first appeared on Earth. It will be seen that CO_2 (dark blue) has fallen steeply from 5,800 ppm to 417 ppm. Its density has no correlation with temperature (pale blue). It is now even lower than during the Permian extinction.

In 2009 the US Environment Protection Agency – a leading manufacturer of global warming propaganda - declared that CO_2 is a “pollutant” and posed a "danger" to human health and welfare requiring that it be regulated under the US Clean Air Act of 1970.

This is grossly misleading. We exhale over 2 lbs (I kilogram) of CO_2 each day. So do large animals. By photosynthesis of CO_2 and sunlight, in conjunction with water and the enzyme Rubisco, the energy of the Sun is converted into food for plants including all trees. This is a critical exchange between an atmospheric gas and the Earth’s biosphere. It is this process that also creates all of the oxygen that we and all other creatures breathe. It is released by conversion of H_2O, in combination with carbon, into energy.

It is rising CO_2 that is responsible for the greening of sub-Saharan Africa and for an 80% - 100% rise in crop yields in Western Europe since 1960 (Figures 42 and 43 at p 101 and related text) .

If it falls to 150ppm all life ceases.

False Scares and Alarms

Many will be familiar with some of the scares and alarms routinely put into widespread circulation to sustain the delusion of global warming dogma. They include the ‘great walrus suicide leap’, the ‘great starvation of the polar bear’, the ‘great Pacific plastic island’, the ‘great demise of the Beobab tree’ and the ‘great decease of the Great Barrier Reef’.

Many of these entirely false accounts of disaster are graphically described and exposed in a recent work by the co-founder of Greenpeace[64]. He resigned from Greenpeace when it became complicit in abusing science and manipulating evidence for political ends.

He sets out how the fraudulent IPCC propaganda of CO_2 induced climate change has been held responsible for higher temperatures, lower temperatures, devastating blizzards, severe rise in sea levels, mass species extinctions, total loss of ice at North Pole, dying forests, death of coral reefs, drought, fire, floods, fatal heat waves, crop failure and food shortages, acidic oceans that will kill marine life, millions of climate refugees, intensified hurricanes, increased cancer, cardiovascular disease and mental illness, thinner pigs and fatter horses.

None of these is sustained by empirical evidence. Atmospheric CO_2 has no causal connection with any of these manufactured scares and reckless predictions[65].

---

[64] Dr Patrick Moore *“Fake Invisible Catastrophes and Threats of Doom”* Ecosense Environmental Inc 2021.pp30 - 34

[65] See Nieboer J. Appendix to *“CLIMATE All is Well All will be Well”* Bruges Group October 2021 Amazon.

II. Fear

Timendi causa est nescire (Seneca)
Ignorance is the cause of fear

Extinctions

Fear and the use of intimidation are means by which fundamental shifts in opinion of whole communities may be procured. They create a rising sense of anxiety and insecurity which informs not just individuals but engenders a collective condition of latent alarm.

The virus of IPCC dogma has infected thousands with the belief that CO_2 induced warming will cause mass extinctions. As will be seen from Figure 11 since multicellular life first emerged atmospheric CO_2 has declined inexorably from concentration of 5,800 ppm to 280 ppm by 1900 AD when it began to rise slightly with human emissions to 417ppm. During Ice Ages it has fallen to 180ppm.

Mass extinctions occur at times of low density of CO_2 which stultifies plant growth due to collapse in photosynthetic activity. The decline in CO_2 in the Carboniferous period depicted in Figure 11 came with the proliferation of vegetation its consumption of CO_2 by photosynthesis. The decline was exacerbated by the steep fall in temperature in the early Permian period causing vast absorption of the gas by oceans. These factors are believed to have led to the Permian extinction of 95% of marine species and 70% of terrestrial species 275 million years ago.

At CO_2 levels of 200ppm – only 80ppm below pre-industrial levels - photosynthesis falters and causes severe reductions in plant growth and reproductivity. Biomass collapse is as much as 70%.[66] At 150 ppm all photosynthesis ceases and life become extinct.

Increasing levels of atmospheric CO_2 – even to as little as the current 417ppm – have effected not only profound improvements in crop yields but also the greening of vast regions of dry lands since 1960 in Africa and China. There are two reasons for this. Increased CO_2 itself promotes photosynthesis. In addition the increase in CO_2 also results in plants closing their stomata[67] to adjust to more of the gas thus drastically diminishing water loss through these openings.

The optimum levels of atmospheric CO_2 for plant growth are 1250 – 2000 ppm – 3 to 5 times current density. Humanity and the entire biosphere would then secure highly beneficial increases in food yields and optimum conditions for species to thrive without any risk of global warming.

It is CO_2 that ensures preservation of species – not their extinction.

Fear of Scarcity

Not having sufficient to meet the demands of existence has been and remains an ineradicable deep human anxiety.

---

[66] Gerhard LM and Ward JK *Plant responses to low level CO_2 of the past* 2010 Ne Phytologist 188 P.691.
[67] Minute orifices in the underside of leaves which are portals for incoming CO_2 and outgoing oxygen.

Fear has been the subject of much research and enquiry. It has been found[68] that it causes a myopic inability to comprehend existing situations or the way actions they will affect them in the future. It directly impairs cognitive function – essential multiple mental abilities, including learning, thinking, reasoning, remembering, problem solving, decision making and attention. This in turn degrades the quality of decisions and choices and thus exacerbates the condition giving rise to the fear.

The phenomenon of fear of scarcity has dominated human responses to the rise in population that began with the growth of the open market economy. It was that which impelled Malthus, Jevons, Vogt, Erlich, the Club of Rome[69] to make their predictions of existential peril.[70]

Members of the so-called Club of Rome have variously asserted that "*World population needs to be decreased by 50%*"[71] - "*In order to stabilise world population we must eliminate 350,000 a day*"[72] - "*We must shift our efforts....to the cutting out of the cancer of [uncontrolled multiplication of people]. The operation will demand many apparently brutal and heartless decisions*"[73].

Yet, as is described in Part 7, the world's population has risen since 1960 from 3billion to 8billion. Notwithstanding such a vast increase, in all open market economies food supplies came into surplus and extreme poverty was virtually eliminated[74].

Fear of the unknown

It is well established that fear is a positive condition when it is matched to a known given threat. The sudden raising of alertness that comes with acute cognisance of immediate risk commonly supplants thought with an instant and appropriate reaction[75].

However such optimal response depends on clear perception of the material constituents of the risk itself. It is the experience of many car drivers that they have, as it were by instinct, avoided accidents by immediate and spontaneous action and clear perception of all that is happening. Such is possible only with comprehension of the factors of imminent danger. It is an instant experience of the present moment – both 'momentous' and 'instantaneous'.

The fears conjured up by the theories of cancer proliferation, acid rain, nuclear winter and ozone 'hole'[76] were dissipated completely by the disclosure of the contradictory empirical evidence of science in each case. That this was possible was due to the fact that these false alarms did not depend on predictions but were based on false assertions about known existing conditions.

When actual risks are unknown, however, then fear cannot be abated by such means. It has been found that without known and certain parameters to envisaged risk there is no means of

---

[68] Jiaying Z. et al *Psychological Responses to Scarcity* 26 February 2018 Oxford University Press https://doi.org/10.1093/acrefore/9780190236557.013.41.
[69] Describes itself as an 'invisible college' of some 70 'experts'. They have included all of the principal adherents to the dogma of global warming.
[70] See Part 4 for a full examination of these predictions.
[71] Henry Kissinger Former US National Security Adviser, US Secretary of State and member of Club of Rome.
[72] Jacques Cousteau French naval officer, oceanographer, filmmaker and author member of Club of Rome.
[73] Erlich P. Professor of Biology Stanford University *'The Population Bomb'* 1968 Sierra Club Ballantine Books.
[74] See Part 7 which consists of a comprehensive review of humanity's progress since the coming of the English industrial revolution.
[75] Pierre J. M., M.D. *'How does fear influence risk assessment and decision making'* Professor Dept of Psychiatry and Behavioural Sciences University of California, psychologytoday.com/us/blog/psych-unseen Posted July 15, 2020.
[76] See Part 4 pp48 – 52 for a comprehensive explanation of these alarms.

confining or rationalising the insecurity that it induces. Research studies[77] confirm what is obvious - the nature of uncertain threat is that it is unpredictable. The threat cannot be reduced to the confines of probability, timing, intensity or duration. A predictable threat in the reality of the present moment can be resolved by the fight or flee response but it subsides once the threat disappears or there is a resolution of the potential outcome.

The fear of the unknown and the insecurity and fear it creates cannot be similarly dissipated. It fades only when the threat is not renewed over time. It is for this reason that the IPCC repeats and ratchets its predictions about how near we are getting to the abyss in order to maintain intimidation.

Fabrication of disasters

It has been established that *"Feelings of dread are the major determiner of public perception"*.[78] There seems to be no limit to the extent of deceit involved in falsifying of 'evidence' to justify the lie about global warming. The co-founder of Greenpeace himself has given an account of some of the worst fabrications of natural disasters[79].

It has been shown that fear paralyses efforts to think clearly about the underlying risk. It is common for fear to distort and misperceive the extent of the risk. In particular it induces a disregarding of the probability of the occurrence of the event or circumstance giving rise to fear.[80] Moreover, since environmental risks are almost invariably perceived to have been imposed by others, they excite indignation and blame and thus further excess reaction. These characterise fear of all kinds but are particularly important when considering two sources of insecurity and alarm found deep in the human psyche – fear of scarcity and fear of the unknowable.

The IPCC maintains its intimidation by the very fact of maintaining the dogma of global warming as a prediction. It does not call into aid actual scientific observation, testing and measurement of the processes of spectroscopy that govern CO_2 absorption of infra-red long wave radiation from the surface of the Earth which is at the very core of the greenhouse effect.

The IPCC knows that application of the scientific method would reveal 'saturation' of CO_2 that renders it impossible for rising density to cause global warming. It resorts to and depends upon claims of catastrophe, intimidation and fraudulent assertions of an overwhelming 'consensus'. [81]

The most recent outburst of the UN Secretary General at the so-called Conference of the Parties of November 2022 in Egypt is no more than a repetition of the countless forecasts of imminent doom that it has made since 1990 – *"We are headed for a global catastrophe"*[82]. In 2009 the leading advocate of the cult of climate change warned the President Obama had *"only 4 years to save the planet*[83]*"*. Hansen testified, falsely, to US Congress on 23 June 1988 that it was 99%

---

[77] Gorka S. clinical psychologist University of Illinois at Chicago College of Medicine *'Fear of the unknown common to many anxiety disorders'* Journal of Abnormal November 18 2016.

[78] Slovic P. et al. *Risk as analysis and risk as feelings: some thoughts about affect, reason, risk, and rationality. Risk Analysis* 2004; 24:311-322.

[79] Patrick Moore Op Cit.

[80] Slovic P, Finucane ML, Peters E, et al Op cit,

[81] *"Unless we announce disasters no one will listen."* Houghton J. *"Global Warming: The Complete Briefing"*. Lion Press 1994 Chairman of the Intergovernmental Panel on Climate Change's (IPCC) scientific assessment working group,

[82] Opening statement of Antonio Guterres November 9 2022.

[83] Dr James Hansen director of The Goddard Institute for Space Sciences until 2013.

certain that dangerous global warming was caused by the building up of carbon dioxide in the atmosphere[84]. Just 4 days later a UN conference on the atmosphere declared that the ultimate consequences of climate change for humanity were *"second only to a global nuclear war"*[85].

Impact on children

The impact of such terrifying propaganda on the young has now reached pathological levels. Children are not only taught the appalling falsehood of global warming and its terrors for humanity. They are also conscripted to protest against it[86] in schools.

A recent survey reported in the Lancet[87] to evaluate distress, functioning and negative responses of 10,000 children and young people to predictions of climate change found that 59% were very or extremely worried. Over 50% reported anxiety, anger, powerlessness, helplessness and guilt. More than 45% of them said their feelings affected their daily life and functioning. A high proportion (75%) found the future frightening and even more (83%) felt betrayed by the perception that humans had failed to take care of the planet.

The syndrome of fear is ruthlessly exploited to maintain the dogma and suppress dissent.

The shocking complicity of the broadcaster Attenborough in the fraudulent depiction of walruses being driven to suicide by lack of sea ice would have continued traumatising women and children had it not been painstakingly exposed 3 years later. Netflix and WWF working with the BBC produced a feature in "Our Planet" on 5 April 2019 of film footage intended *'to elicit an intense emotional reaction: unforgettable nightmare inducing horror'*. Dr Susan Crockford[88] sets out in devastating detail this manipulation of evidence for eco-socialist purposes. She relates the terrified shock suffered by children witnessing these heavy, helpless creatures being bounced off crags and boulders as their burst or dismembered bodies fall in death agony on the beach 80 metres below. These scenes were in reality occasioned by polar bears stalking walrus massed in great numbers of normal 'haulouts' on a sea cliff at Cape Schmidt of northeast Russia.

It is open to anyone with internet access - and whose doubts compel them - to discover the simple truth. The CO_2 greenhouse effect cannot cause global warming and has never done so. Disabling of ignorance has never been easier.

III. Guilt

Propaganda and the "Enemy"

The imposition of guilt upon the inhabitants of the developed world as a tool of climate catastrophe indoctrination has been very effective in ensuring its success.

---

[84] See Part 5 p61.

[85] The Toronto Conference on the Changing Atmosphere: Implications for Global Security, 27 to 30 June 1988, issued a warning that humans had unintentionally triggered uncontrolled changes to the atmosphere.

[86] Levif.be/international/la-fanatisation-de-l'enfance-une-modalite-de-l'abus? »

[87] Hickman H MSc et al The Lancet vol 5 no 12 December1 2021. The survey covered 10 countries.

[88] Dr Susan Crockford a Zoologist was from 2004 to 2019 Professor at the University of Victoria Canada. Her book "Fallen Icon" 2022 exposed the charade of falsehood that was the Attenborough 'walrus suicide' atrocity. Dr Crockford's 15 year tenure at the University of Victoria terminated upon her disclosing the falsehood of polar bear starvation due to' lack of sea ice' See *"State of the Polar Bear Report 2020* "Global Warming Policy Foundation Report 48. Wikipedia contains disgracefully distorted and malevolent statements intended to depreciate falsely the professional status of Dr Crockford.

Hayek in his seminal work *"The Road to Serfdom"*, published in 1944 at a time when totalitarian tyrannies dominated Europe and Russia, describes a critical phenomenon that assists the formation of such regimes[89]. He explains that in seeking to weld together a closely coherent and homogeneous body of supporters an important negative element is required. Hayek shows how it is almost a law of human nature that it is easier for people to agree on a negative programme – on the hatred of an enemy – or the envy of the better off – than on any positive task.

For those who seek to indoctrinate masses with a dogma - and secure their unreserved submission to it - the notion of the 'enemy' is an essential tenet of the creed[90] and is indispensable to gaining acceptance of drastic change. It is disturbing to note that, as applied to the rise to dominion of the dogma of global warming, Hayek's diagnosis has been proved correct in all essential respects.

Mankind as an enemy and a cancer

In 1972 the so-called Club of Rome published its discredited 'Limits to Growth' with modelled predictions of disaster due to population growth and exhaustion of resources. In 1974 its next 'Report' called for the re-structuring of human society and a single global economic system[91]. The book cast mankind as a malignant tumour[92] - *"The World has Cancer and the Cancer is Man"*.

The first nuclear arms limitation treaty in 1972[93] had relieved human terror of nuclear war and of the Soviet Union as the traditional enemy. It opened an era of condemnations of humanity as the enemy of Nature all of which were falsified as set out in Part 5 below. In 1991, the Club published 'The First Global Revolution' urging the creation of an image of Man as essential to the propaganda of global warming gaining a hold on opinion and policy. The book noted that ".. *the sudden absence of traditional adversaries has left governments and public opinion with a great void to fill. New enemies have to be identified, new strategies imagined, and new weapons devised"*. It also noted that historically, social or political unity had commonly been motivated by identifying enemies in common just as had been described by Hayek some 30 years earlier.

"*The need for enemies seems to be a common historical factor... In searching for a new enemy to unite us, we came up with the idea that pollution, the threat of global warming, water shortages, famine and the like would fit the bill....All these dangers are caused by human intervention and it is only through changed attitudes and behaviour can they be overcome. The real enemy then is humanity itself."*[94]

The end of the Cold War was declared in 1989[95] following the fall of the Berlin Wall. It allowed the accumulated human insecurity to be preyed upon to instil both fear of the coming disaster of global warming and also guilt for creating it.

---

[89] Hayek's analysis of socialism and freedom forms the subject of Part 6 below. It includes a section on Propaganda.

[90] *Hayek Road to Serfdom* 1944 George Routledge & Sons p143. Professor Hayek was an economist, legal theorist and philosopher celebrated for his defence of classical liberalism He shared the 1974 Nobel Memorial Prize in Economic Sciences due in part to his work showing how changing prices communicate information.

[91] Mesarovic C. et al *The Limits of Growth; Mankind at the Turning Point* Chs 4-9 1973 edn. 1972 *'Limits to Growth'* sis discussed in Part 4.

[92] Mesarovic C. et al Op cit *"The World has Cancer and the Cancer is Man"* was quoted in chapter 1 of the first edition.

[93] SALT I is considered the crowning achievement of the Nixon-Kissinger strategy of détente. The ABM Treaty limited strategic missile defences to 200 interceptors each and two missile defence sites, one to protect the national capital, the other to protect one ICBM field.

[94] *The First Global Revolution- A Report by the Council of the Club of Rome* – Alexander King, Bertrand Schneider – - Random House. Inc Pantheon Books (1991) 1993 edition .Op Cit pp 70 and 115.

[95] Declared at the Malta Summit US President George Bush and Soviet General Secretary M Gorbachov December 1989.

The existence of an underlying and abiding sense of guilt for the supposed damage to the natural balance of the planet has been a dominant factor in the suppression of contradictory evidence and opinion as to its existence or cause. Attenborough has added his own paeon of hate describing humans as a "plague on the Earth" in a 2013 interview with the Radio Times[96].

It is for this reason that doubt and contradiction are condemned as immoral[97] and a crime[98]. Those who have the courage and integrity to speak out are vilified and ridiculed[99], their careers put in jeopardy and their voice silenced.

Collective guilt and reparations

The phenomenon of collective guilt in the context of conceived blame for 'climate change' has been the subject of close study and research.[100] It has been found that collective guilt is a potent force in conditioning behaviour and attitudes. It requires the existence of a collective identity, a presumption of responsibility for harm and a perception of the illegitimacy of harm.

The effect of feeling included in a collective category of guilt has a notable effect on willingness to respond to demand for action to combat it – the deeper the sense of guilt the more unconfined is the desire for fundamental change.

A consequence of collective guilt for climate change is a *'general support for reparations'* and specifically by redistribution policies[101]. Moreover, collective guilt is found to extend particularly to predictions of harms. It forms a skein around both unintended harm and also failure to prevent it. Collective guilt was reported to be a much more potent force for impelling demands for drastic change than collective fear.

It is the gospel of the guilt of the West preached by the apostles of climate change dogma that is the justification of demands for reparations for the supposed 'crime' and which accounts for the craven gullibility of those accused who submit to them. It is incomprehensible that any regard is paid to these outbursts of ingratitude and ignorance.

The ascent of developing nations from poverty, disease, and fear that had plagued all human existence, save for the very few, are summarised in Part 7 of this book – the Sunlit Uplands of life in a golden age.

SUMMARY

Global warming due to human CO_2 emissions is the dogma of an unaccountable political bureaucracy – the IPCC. It has no scientific validity. It depends solely on predetermined modelling and falsified records. It is contradicted in all respects by findings of spectroscopy and actual empirical observation. It is maintained by deceit, fear and guilt.

---

[96] Gray, Louise 22 January 2013 *"David Attenborough – Humans are plague on Earth"* Daily Telegraph.

[97] *'Human induced climate change is a moral wrong'*[97] Beinhocker E. Professor of Public Policy Practice Blavatnik School of Government University of Oxford published in "Democracy" June 24 2019.

[98] James Hansen Director of The Goddard Institute for Space Sciences until 2013 was an obsessed believer in the dogma of global warming. He called for CEO's of fossil fuel companies to be put on trial for *"high crimes against humanity and nature"* .

[99] Speech of Lord Lawson Bath University 26 April 2014*"I have never in my life experienced such extremes of personal hostility, vituperation and vilification."* page 82 below footnote. See also the dismissal of Dr Susan Crockford from her 15 year tenure as Professor of the University of Victoria following her courageous exposure of the multiple falsehoods of eco socialist claims that polar bears were starving due to retreating sea ice.

[100] Ferguson M.A. and Branscombe N.R. *'The Social Psychology of Collective Guilt* '2004 Oxford University Press.

[101] Ferguson et al Op cit p 255b.

PART 3

THE COMING OF THE INDUSTRIAL AGE

I. The Great Transformation

Industrialisation and Nature

The preservation of the air, of water, of the land, of eco-systems, of the atmosphere and of species – all of these have been raised by humanity to matters of high concern. That this should be so is a testament to the advances in civilisation that have occurred since the advent of the industrial revolution in England. It is one of the great achievements of human progress.

This Part 4 includes an examination of the origins and development of this concern from the time when the industrial revolution, over just a few generations, began to overturn the conditions of existence prevailing since at least the advent of human settlements.

Part 5 demonstrates how it was that at the end of the 1980s the sincere concern for care of the natural world began to be suborned and then overwhelmed by a form of hatred of humanity itself for what it had created and a rejection of the economic and political structures that were condemned for making it possible. This eco-socialist assault on the free open market economy and on the liberal democracy of nation states[102] made its principal advance in just 4 years from 1988 to 1992. Its weapon was fashioned by the welding of the dogma of global warming on to existing and genuine anxieties as to mankind and the natural world.

The dogma utterly displaced the sincere sense of unease about the natural environment with a form of eco-socialism having as its purpose the reconstituting the economic model of western developed countries and their systems of governance. It was of little importance for this purpose what means should be adopted to achieve this end. "*We came up with the idea that pollution, the threat of global warming, water shortages, famine and the like would fit the bill* [103]."

For Schneider,[104] the leading apostle, it was as morally convenient to publish in 1976 claims of an imminent Ice Age[105] as it was to publish in 1989 warnings of imminent severe global warming - to publish findings that saturation rendered it impossible for CO_2 to cause warming even at 8 times current density[106] and then to become the main scientific advocate for global warming[107].

To appreciate the eco-socialist 'First Global Revolution[108]' which was to overwhelm honest environmental unease as to human impact on nature it is instructive to review the phenomenon of the Industrial Revolution itself - to consider its causes, its perceived consequences and the ground won by it from the poverty, disease, want and fear that had been the lot of humanity save only for the very few.

---

[102] See for example Klein M, social activist, and filmmaker *'This Changes Everything: Capitalism vs the Climate'* 2015 Simon & Schuster.
[103] A Report by the Council of the Club of Rome – Alexander King, Bertrand Schneider – - Random House, Inc Pantheon Books (1991).*'First Global Revolution'* was to be a *'blueprint for the 21st century'*[103]. Launching Press conference Washington DC Sept 16 1991.
[104] See Part 1. Also Part 4.
[105] Schneider S. "The Genesis Strategy".1976 Plenum Press.
[106] Schneider S. "Global Warming Are we entering the greenhouse century?" Sierra Club Books 1989 Vintage books 1990.
[107] *Atmospheric Carbon Dioxide and Aerosols. :Effects of Large Increases on Global Climate'.* SCIENCE. Vol 173 9 July 1971 173 pp138 – 141
[108] The title of "*The First Global Revolution – Report of the Club of Rome*" King and Schneider. Published by Pantheon Books 1991.

The Economic Transformation

The work of empirical analysis by researchers at Oxford University and of Global Change Data [109] has provided graphic illustrations of many of the transformations reviewed in this book. A considerable debt it due to them for the enlightenment they have made possible.

The period of the Industrial Revolution in England was the first time in recorded history during which there was an increase in both population and at the same time a comparable rise in per capita income. The transformation it wrought is illustrated by the following graphs.

They are but a snapshot of the turbulent re-formation of the economy and of civil society that, in its first and most decisive phase, spanned the period of less than 50 years from 1800 to 1848 - the year in which working class Chartists[110] marched in London demanding electoral reform and Marx and Engels published their Communist Manifesto.

The primary evidence of prosperity is the share of people in their national wealth.

Figure 12

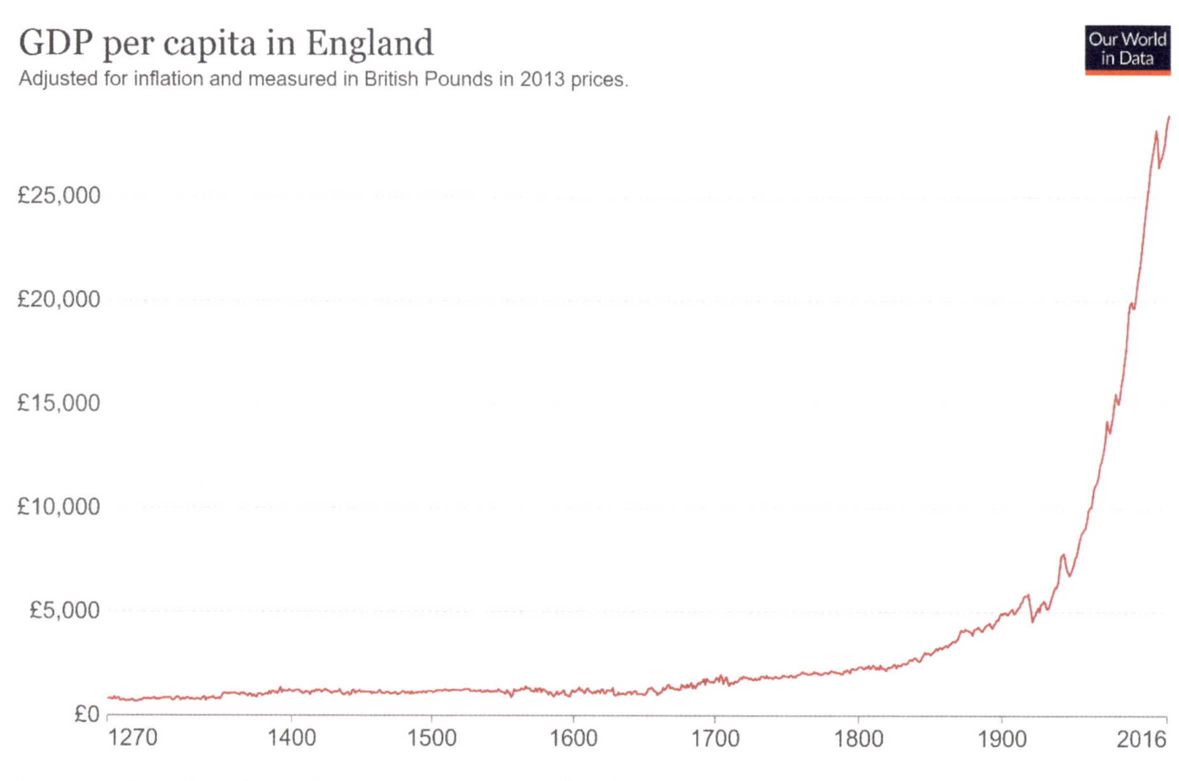

GDP as a whole in England in the period 1820 to 1900 grew 450% from $70.2bn to $312.53bn. This may be compared to France and Germany.

---

[109] Our World in Data.

[110] The People's Charter called for ▪ A vote for every man aged 21 and above. ▪The secret ballot to protect the elector in the exercise of his vote. ▪ No property qualification for MPs , ▪ Payment of Members enabling persons of modest means to attend to the interests of the nation. ▪ Equal constituencies having the same amount of representation for the same number of electors.▪ Annual Parliamentary elections to check incidence of bribery. Modest pleas indeed.

Figure 13

GDP, 1 to 2018

GDP adjusted for price changes over time (inflation) and price differences between countries – it is measured in international-$ in 2011 prices.

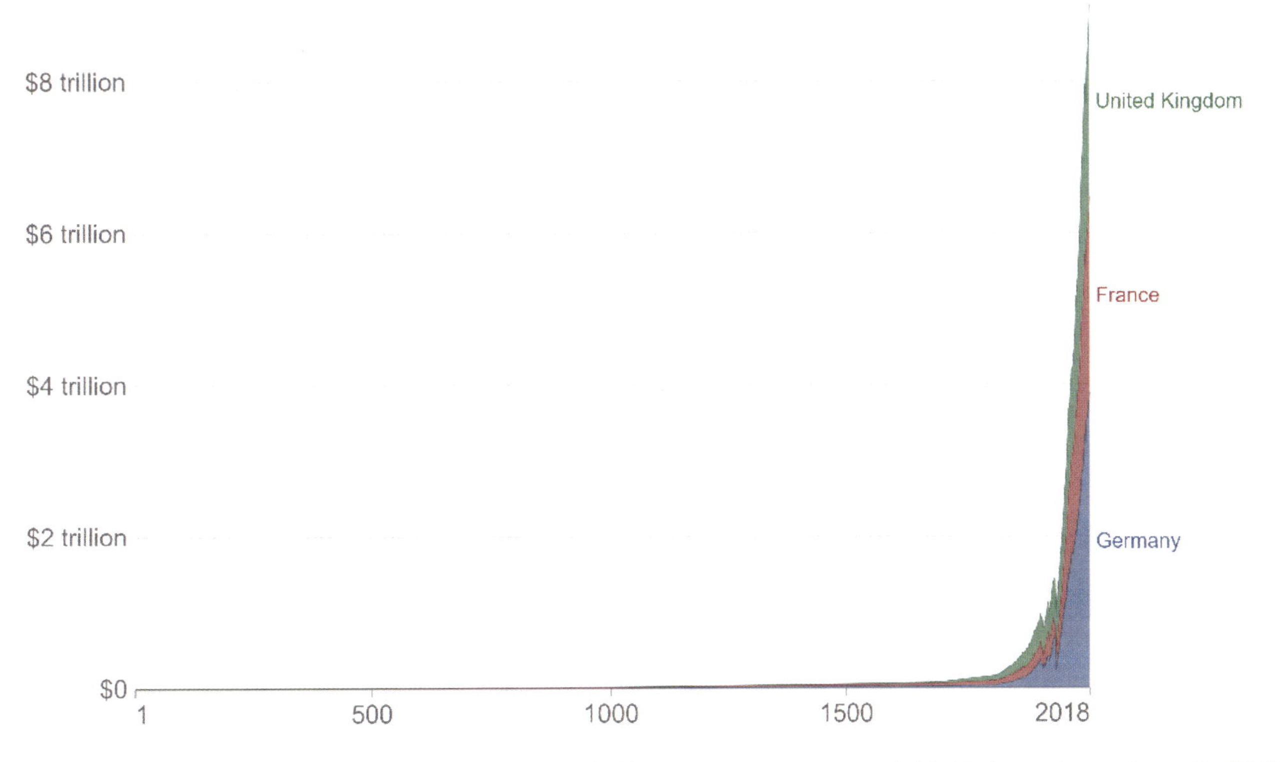

Over the period 1820 to 1900 GDP grew in Germany from $39.15bn to $258.78bn (660%) whilst in France it grew from $56.53bn to $186.10bn (330%).

Humanity has enjoyed just this one fundamental and enduring economic transformation. Moreover it was accomplished in the period of no more than 4 generations. It was not only a total re-formation of the entire economic system: it was also almost instant in its impact.

It is for that reason that it was a shock. Its reverberations engendered many and varied emotions and responses. But its benefits were so comprehensive and widespread that it ranks as the greatest and the most beneficial transformation of the conditions of human life.

The economic demands made by an average person to maintain existence in 1665, just after the Restoration, were substantially the same as those at the time of Magna Carta 450 years before. Average incomes in England between 1215 and 1650, measured by GDP per capita, were £1,051 when measured in today's prices.

Out of this bleak and hopeless existence there emerged a miracle of economic and social improvement. From an average of £1,051 pa incomes rose to over £30,000 p.a. in 2016; a 29-fold increase in prosperity. An average person in the UK in 2022 had an income and produced as much in a fortnight as an average person in 1650 received and produced in an entire year.

Population

Moreover economic growth engendered higher wages and the rise in population itself accelerated total output of the entire UK economy[111].

Figure 14

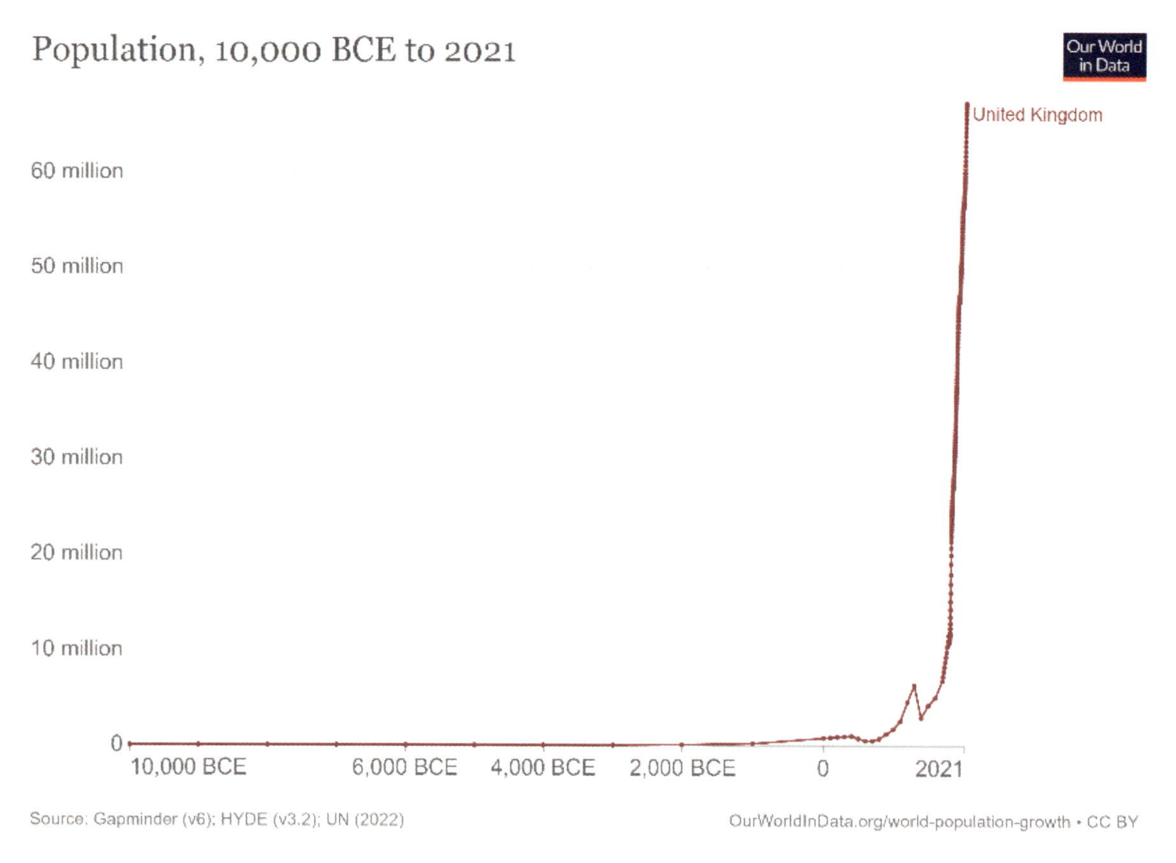

Population in England at the end of the Black Death in the 14th century had been 2.9m. In 1790 it was 12.52m. By 1851 – the year of the Great Exhibition in Hyde Park - it had more than doubled to 27.38m. By 1914 on the outbreak of the World War 1 it had risen since 1790 by nearly 400% to 46.8m.

Yet as has been seen (Figure 12) GDP per capita grew by nearly 700% in the same period.

As is demonstrated in greater detail in Part 7 the predictions of catastrophe as to the impact of population on resources of food were falsified by the very industrial and scientific revolutions that gave rise to it. Mechanisation and land enclosure not only multiplied crop yields but steam power also opened up new domestic markets as well as the vast grain supplies of the US mid-west later in the 19th century.

The manufacture of nitrate fertilisers from atmospheric nitrogen allowed steeply increasing agricultural yields at decreasing costs. The Green Revolution[112] of 1960 – 2000 of high yield disease and wind resistant wheat created food surpluses in countries once beset by famine.

[111] Source: Gapminder (v6); HYDE (v3.2); UN (2022). Our World In Data Population Growth.
[112] See pp 98 – 101. Also Part 4 pp 40 – 43.

II. Origins and causes

The seed beds of change

What made the Industrial Revolution in England possible was a combination of conditions and advantages which, over a period of 4 generations, utterly transformed the conditions of existence which had prevailed for 8,000 generations of humankind.

Foremost among these were:-

- Abundance of energy in the form of accessible coal.
- An open market economy.
- Institutions and systems of credit and investment.
- A liberal democratic system of government.
- Free trade.

Coal

It was coal[113]. that provided the energy for steam and, later in the 19th century, for electricity generation. It is instructive to see the advance of coal extraction in this period.

Figure 15

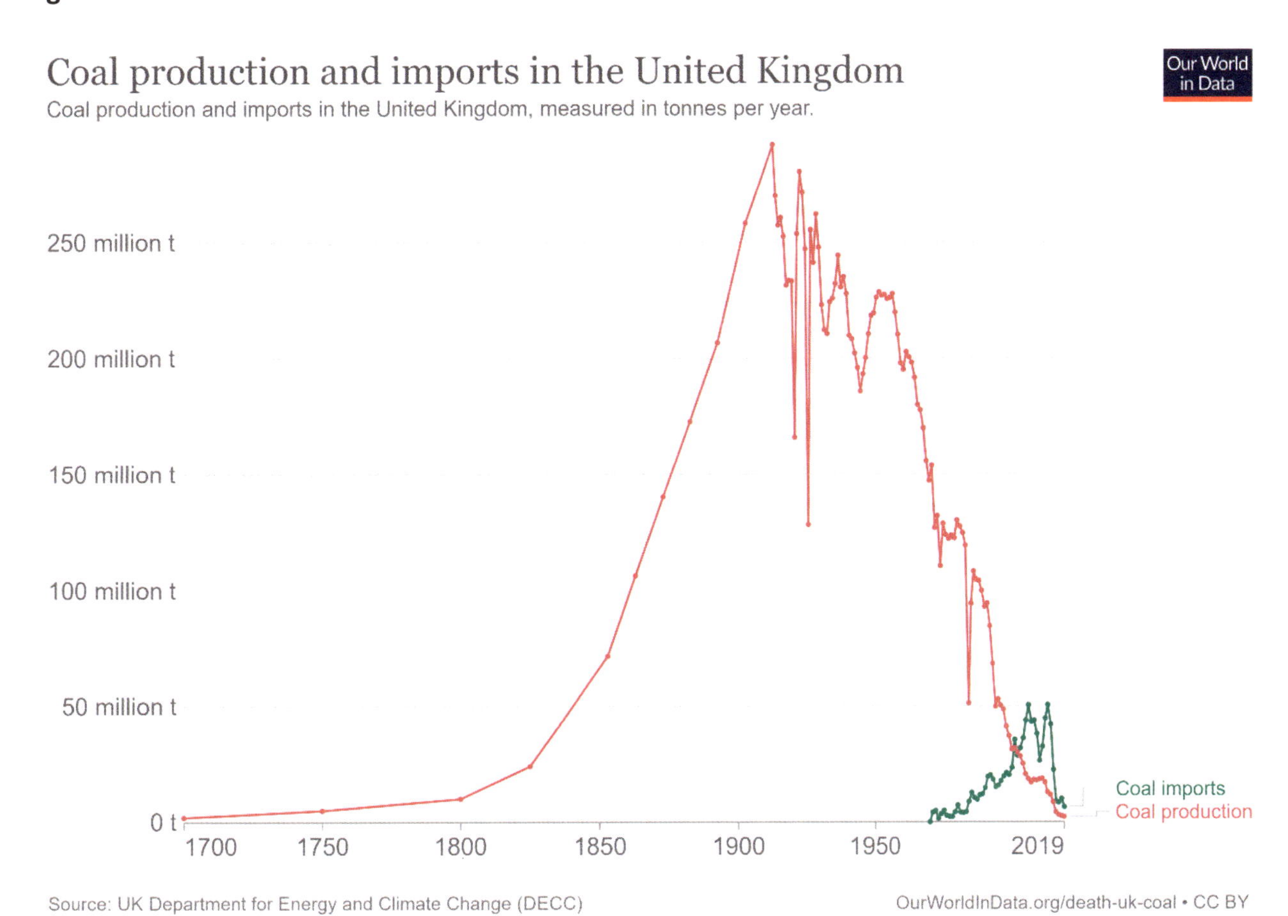

[113]See Darwall R. *The Age of Global Warming* Quartet Books 2013 p 25. "If the combustion of coal and other hydrocarbons had been severely restricted from the start of the 19th century the economic take-off of the Industrial Revolution would not have happened: we would all be a lot poorer, our lives would be shorter and most of us would be earning our living working in the fields".

Coal extraction was very limited prior to the industrial era. Rising demand for it in the late 18th century and early 19th century[114] led to the development of both the stationary and also of the locomotive steam engine using train tracks at collieries.

It was coal and steam that powered the industrial revolution. The United Kingdom still has reserves of identified strata of coal of up to 23 trillion tonnes lying off the north east coast. Its onshore gas resources are estimated to provide for over 300 years of combustion at present rates with just 10% sufficing for 40 years. The United Kingdom remains rich in hydrocarbon fuel resources.

Impact on emissions of CO_2

The impact of the English industrial revolution was in every respect beneficial to humanity. As is set out in Part 4 none of the alarms as to mankind's impact on nature were found to be justified by science. Only a negligible amount of CO_2 was emitted by the UK until 1960 when it remained minute as a proportion of world emissions and these have no adverse effects whatsoever.

Figure 16

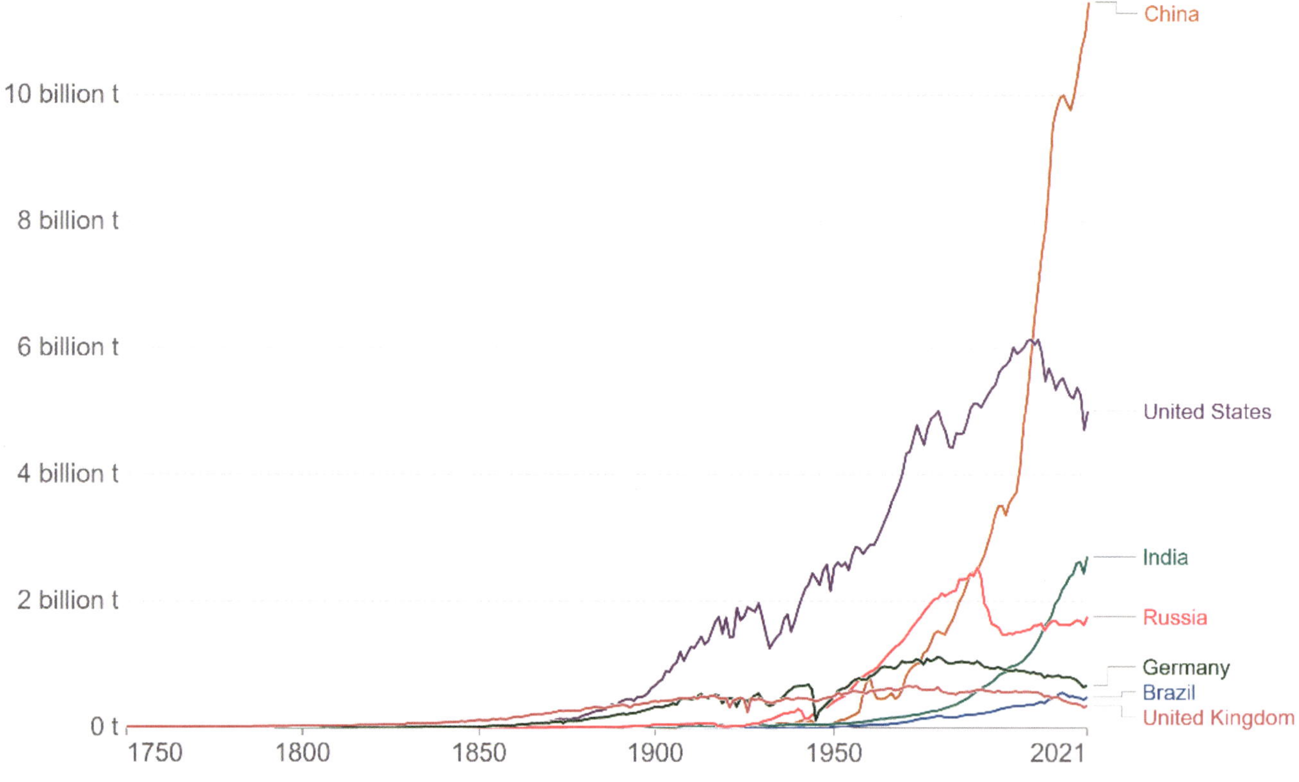

[114] The most famous being by locomotive engines produced by George Stephenson for Killingworth High Pit colliery in Northumberland.

III. Open market economy

Banking and Debt finance

The growth made possible by scientific discovery, a limitless energy store, rising population density and agricultural and technological advances also depended for its extent and variety on the availability of financial credit to fund the new ventures of the industrial revolution. It was this that allowed and promoted the exploitation of the opportunities for human welfare and prosperity that were arising on a scale of hitherto unimagined extent.

Banking enterprises in Great Britain had already been created in the 18th century. They included at least ten of City banks of the 20th century.[115] These were largely funded by anonymous stockholders and conducted by professional managers.

Clearing facilities, security investments, cheques and overdraft protections were introduced. The first overdraft facility was set up in 1728 by the Royal Bank of Scotland. Settlements of cheques were at first made by courier to the issuing bank but by the 19th century a dedicated bankers' clearing house had been established.

City of London merchant-banks of the immigrant families of Rothschild and Baring came to dominate world banking over the 19th century particularly in sovereign debt and international trade. A great impetus to provincial banking resulted in 1797 when Parliament authorised the issue notes of low denomination with suspension of cash payments due to the war with France. This rapidly extended the availability of credit just as the industrial revolution took hold.

Joint stock banks sprang up in provincial towns. In 1826 the Bristol Old bank converted from a private bank to become the first joint stock bank. In that year a new joint stock bank, Lancaster Banking Company, was formed which was quickly followed by the Manchester & Liverpool District Banking Company and other institutions among them the Provincial Banking Corporation which became the National Provincial Bank with twenty branches across England and Wales.[116]

Typical of the spontaneous response to the swift evolution of commercial opportunity was the development of the British Linen Company from a linen manufacturing concern into an extensive banking enterprise. Its charter had included the conduct of banking as an incidental facility of its operations. Its banking services facilitated trade with promissory notes issued to pay creditor agents, weavers, manufacturers and other suppliers before itself issuing true bank notes, payable on demand but without interest.

The transition from a commercial linen trading enterprise to a banking network was made by stages each in response to demand and to economic conditions. In 1800 there were 18 branches of the bank with a further 25 added by 1850 and no less than 70 new openings by 1900.

[115] 1664 Child & Co; 1664 C Hoare Co; 1690 Barclays; 1692 Coutts; 1719 Drummonds; 1765 Lloyds' 1786 Hottinger; 1804 Schroder; 1810 Brown Shipley; 1816 Cater Allen.

[116] Provincial Banking Corporation was formed from Day, Nicholson and Stone, the Bank of Wales in 1864/5 and Fincham and Simpson in 1871. It was reorganised in 1870 and became the London and Provincial Bank Limited being taken over by Barclays Bank in 1918.

Joint Stock Companies

The law of contract had been developed throughout the 18th century in the common law courts and also from the early 19th century in the courts of equity. Security of ownership of land was afforded by forms of tenure upheld by law including parliamentary statutes. Title to rights to inventions was protected by patent registration.

The vehicle for industrial investment evolved from partnerships or informal joint ventures to become the joint stock company that offered shares for subscription by investors able to transfer them for value on a public exchange. The perfecting of this entity for financial investment in risk ventures came with the creation of limited liability of its members.

The earliest joint-stock company to be recognised in England was the Company of Merchant Adventurers to New Lands, founded in 1551 with 240 shareholders. Like most early ventures it depended on monopoly not upon competition. Such was the basis of the South Sea Company.

The great advance in the funding of enterprises came with issue of shares in limited companies traded on a public exchange. In 1602 the Dutch East India Company had issued shares which were admitted to trading on the Amsterdam Stock Exchange. Moreover it was the first incorporated entity conducting international trade. In 1612, it became the only such corporation combining 'locked in' capital with limited liability.[117]

The scope of incorporation in Great Britain was limited by the need for a royal charter or a private Act of Parliament. However with the coming of the railway booms of the 1830s and of the 1840s Parliament passed the Joint Stock Companies Act 1844 which enabled a company to stipulate in its Articles that it had limited liability. Full limited liability without such process was conferred in 1855.

The modern limited liability company offering tradeable shares to members of the public on a public investment exchange had arrived. It was the means by which credit could be extended and risk assumed. It carried forward the tide of investment and prosperity from thereon.

Credit and Risk. Equity finance

Before the onset of the industrial revolution the London Stock Exchange[118] traded mainly in government securities. Government sponsored or chartered entities were created as a form of partnership of government with private investors. These included the Bank of England - also founded in the same year - the East India Company and the infamous South Sea Company.

The coming of the industrial revolution created an explosion of investment opportunities. In particular it fed railway mania as it took hold from 1830 to 1837. By 1836, during the first railway boom, stock exchanges had been established in Liverpool and Manchester. It was those cities that had been linked by railway in 1830 just 12 months before the first Reform Bill was presented in the House of Commons. The second railway boom of 1844-45 spawned stock exchanges in

---

[117] See further Harris R. *'A new understanding of the history of limited liability: an invitation for theoretical reframing'* Published online by Cambridge University Press 8 June 2020 Journal of Institutional Economics Volume 16 , Issue 5 , October 2020 , pp. 643 –664.

[118] Founded in 1694 in the same year as the Bank of England.

Glasgow, Aberdeen, Edinburgh and in English towns and cities serving investors whose funds derived from local and regional industries. Provincial stock exchanges were thereafter established in most major industrial conurbations including Birmingham, Bradford, Bristol, Halifax, Huddersfield, Hull, Leeds, Leicester, Newcastle, Nottingham, Sheffield and York. Opening of exchanges for securities of joint-stock companies proliferated from 1875[119].

The London Stock Exchange dominated the market for domestic and international government debt securities, whilst the provincial exchanges were the principal markets for the securities of small regional companies. They were the investment arteries of the body industrial.

IV Middle Class Democracy

The Peterloo massacre of 16 August 1819 in St Peters Field Manchester marked an historic confluence of impulses towards the emancipated political and economic systems that fostered the advances of the 19th century in Great Britain.

A serious slump had followed the end of the war with France in June 1815. It was aggravated by the failure of the harvest[120]. Added to the distress of severe unemployment was the suffering caused by the high and rising price of bread due to the Corn Laws imposed that year just 3 months before the Battle of Waterloo. These laws created tariffs on imported grain to protect landed owners from cheaper imports. The plight of the poor and the deemed oppression by the landed class were lampooned by George Cruikshank in his famous indictment of protection.

[119] Local exchanges were established in Oldham (1875), Dundee (1879), Cork (1886), Belfast (1897), Cardiff (1892), Halifax (1896), Greenock (1888), Huddersfield (1899), Bradford (1899), Swansea (1903), Nottingham (1909), and Newport (1916).

[120] This itself was occasioned by the loss of solar radiation due to the eruption of Tambora in April that year – the most catastrophic in recorded history.

Moreover there was growing and profound discontent with the absence of electoral representation in Parliament of the rising new middle class and of working people. Only 11% of adult males had the vote. Few of the new industrial towns were included in the electorate.

In 1819, Lancashire had only 2 county MPs and just 12 MPs from boroughs with only 17,000 voters in a population of nearly one million. The major urban centres of Manchester, Salford, Bolton, Blackburn, Rochdale, Oldham and Stockport had no representation whatever in Parliament yet those owning property in more than one constituency could vote multiple times.

Public protest at both the Corn Laws and lack of representation erupted in Manchester following rejection of petitions for manhood suffrage. A public rally in St Peters Field in August 1819 was cruelly dispersed by cavalry charges with loss of life. The 'Peterloo Massacre' shocked the nation.

It was fear of severe civil unrest on the scale witnessed in the Bristol riots in response to the rejection of the 1830 Reform Bill by the House of Lords, that impelled the passing of the Reform Act of 1832. In addition, a larger representation was thought necessary to counter a perceived threat of catholic members entering Parliament following the abolition of laws excluding them as MPs. The non-conformist cities of the Midlands and North were thus admitted to the electorate.

The 1832 Act launched modern democracy in Great Britain.

It was a middle class advance, not of the people generally. The alliance of the working class with the middle class was ruptured for the time being. However the new political structure was essential to the flowering of the industrial revolution. The passing of Acts of Parliament, all within just 25 years, authorising the birth and growth of the railways[121], the creation of the incorporated limited liability joint stock company[122], the creation of effective protections for inventions by patents[123], the passing of the repeal of the Corn Laws and the opening of the long period of international free trade championed by Cobden and Bright were all its beneficial consequences.

Social legislation

Parliament became susceptible to public opinion to an extent unknown save in times of civil war or unrest. This extended to care for those exploited by employers. It was the new political class, inspired by an aristocrat, that introduced fundamental social legislation to give protection from abuses of freedom particularly as they affected women and children. The prime mover of the legislation to improve the conditions of workers in factories was the 7th Earl of Shaftesbury.

In the few years between 1832 and 1848 no less than 7 Acts were passed regulating inspection of factories and prescribing limits as to the employment women, children and young people. The 1844 Factories Act required that children should not work more than 6 hours and that at least 3

[121] A new railway needed an Act of Parliament at a cost of the order of £200,000. During the boom year 1846 alone 263 Acts of Parliament for setting up new railway companies were passed, with the proposed routes totalling 9,500 miles . he Liverpool and Manchester railway cost £637,000 (£55,210,000 adjusted for 2015)

[122] Incorporation of companies generally was introduced by the Joint Stock Companies Act 1844 Companies so incorporated did not have limited liability but could provide for this in the Articles. The Joint Stock Companies Act 1856 provided for limited liability for all joint-stock companies provided, among other things, that they included the word "limited" in their company name.

[123] The Patent Law Amendment Act of 1852 established the modern Patent Office with effective procedures and protection.

hours of education be provided[124]. The 10 Hours Act of 1847 applied prohibitions to those aged 13 – 18 in addition to existing restrictions.

As leader of the Factory Reform Movement in the House of Commons Shaftesbury himself was ultimately responsible for the passage of nearly every labour reform bill from the time that he entered Parliament in 1826. He was noted particularly for the Factory Act of 1847. Marx in his Communist Manifesto applauded the 1847 Act as a victory for the working class. It was also a triumph for the aristocracy.

Prompted by a cholera outbreak in London Parliament passed the Public Health Act of 1848 establishing a central Board of Health. By 1853 over 160 towns and cities had set up local boards of health. Disraeli had committed the Conservatives to social reform and on taking office a Public Health Act was passed in 1875. Among its measures were requirements that all new residential construction must include running water and an internal drainage system and for each public health authority to appoint a medical officer and a sanitary inspector.

It was the 1832 Reform Act that was decisive in bringing about the rule of liberal democracy in Great Britain. It was this that enabled its free market economy to become rooted and to thrive while its abuses were brought under the oversight of the liberal democracy it had created.

V. Free Trade

The concept of trade without levy or tariffs on exchange of goods stood in flat contradiction of the established policy of protection of trade that dictated restrictions, controls and prohibitions imposed to diminish or eliminate the impact of foreign imports on domestic production. Free trade represented the open free market as applied to international trade.

Protection

In the 18th century the price paid by the Great Britain for the imposition of mercantilism on trade was the loss of its American colonies. Trade in colonial America was regulated through the Acts of Trade and Navigation. The Prohibitory Act of 1775, severed all trade between the 13 colonies and England. In April 1776 the Continental Congress of the American colonies repudiated the prohibition declaring that American ports would henceforth be open to all foreign trade. The colonies declared independent political sovereignty just 3 months later.

In the same year as the American colonies cast over economic and political dominion Adam Smith set out with simple clarity the paramount gain to be derived from free trade.[125]

> *In every country it always is and must be the interest of the great body of the people to buy whatever they want of those who sell it cheapest. The proposition is so very manifest that it seems ridiculous to take any pains to prove it; nor could it ever have been called in question had not the interested sophistry of merchants and manufacturers confounded the common sense of mankind.*

[124] The Mines and Collieries Act of 1842 prohibited the employment of women and young children under the age of 10 in mines.

[125] Adam Smith 1776 *Wealth of Nations*, Chapter 3, titled "On the extraordinary Restraints upon the Importation of Goods of almost all kinds from those Countries with which the Balance is supposed to be disadvantageous". Adam Smith favoured the withdrawal from the American colonies.

Repeal of Corn Laws

Just 2 years after the Corn Laws were imposed - and 2 years before the Peterloo Massacre - David Ricardo in his *' On the Principles of Political Economy and Taxation'* (1817) had urged the adoption of free trade on the basis of comparative advantage. He argued that removal of tariffs would lead to a net welfare gain – the gain of consumers outweighing the loss of producers[126] - with a general enhancement of trade. In Britain, free trade was adopted as a paramount principle with the successful struggle for repeal of the Corn Laws in 1846. Protection of the domestic agricultural market had caused severe hardship. The repeal was a watershed in freedom and choice applied to the commerce of nations. The first free trade agreement with France, the Cobden-Chevalier Treaty of 1860 prompted a series of European free trade treaties.

Figure 17

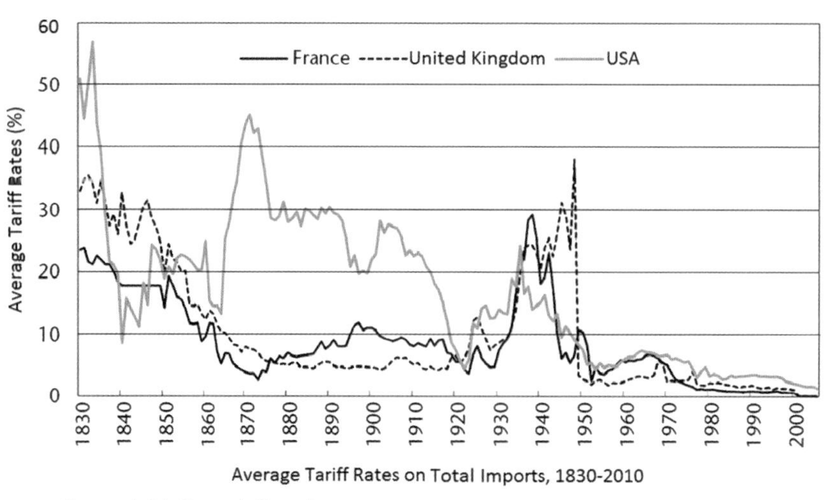

Average Tariff Rates on Total Imports, 1830-2010

Sources: Imlah, Economic Elements

The vice of mercantilism consists of the falsehood that a phenomenon which is essentially variable and dynamic is fixed and determined. It is concerned with the transfer of wealth at the expense of others and not with the creation of wealth. This conception has distorted human response to economic change. As Milton Friedman has noted *'most economic fallacies derive from the tendency to assume that there is a fixed pie, that one party can gain only at the expense of another. The most important single central fact about a free market is that no exchange takes place unless both parties benefit'*.

From Heraclitus[127] to Hayek[128] it has been seen that Nature is change and that it is possessed of the capacity to adapt, evolve and multiply. These are the very qualities of humanity that govern the creation of wealth. In modern parlance it is not at all a "zero -sum" process.

Monopoly and Competition

The vast expansion of human discovery, ingenuity and creativity that characterised the advent of open market economy was an assertion of freedom from a rigid pre-determined view of human

[126] *'Under a system of perfectly free commerce, each country naturally devotes its capital and labour to such employments as are most beneficial to each. This pursuit of individual advantage is admirably connected with the universal good of the whole'.*

[127]. *'You could not step twice into the same river' 'All is flux, nothing stays still' πάντα χωρεῖ καὶ οὐδὲν μένε.*

[128] Hayek FA *'The mind cannot foresee its own advance'.*

commerce. It rested on a 'spontaneously self equilibrating' system. Its consequences for human prosperity and welfare flowed from the optimism and energy imparted by casting off fixed conceptions of wealth creation and a willingness to embrace the new. The industrial revolution impelled the replacement of monopoly with competition and protection with free trade. The upward spiral of population, prosperity, demand and production liberalised international trade. In 1801 the East India Company had accounted for half of the world's trade[129] with sales of £7,602,041[130] yet within 12 years its monopoly had been revoked altogether.

International trade

Figure 18

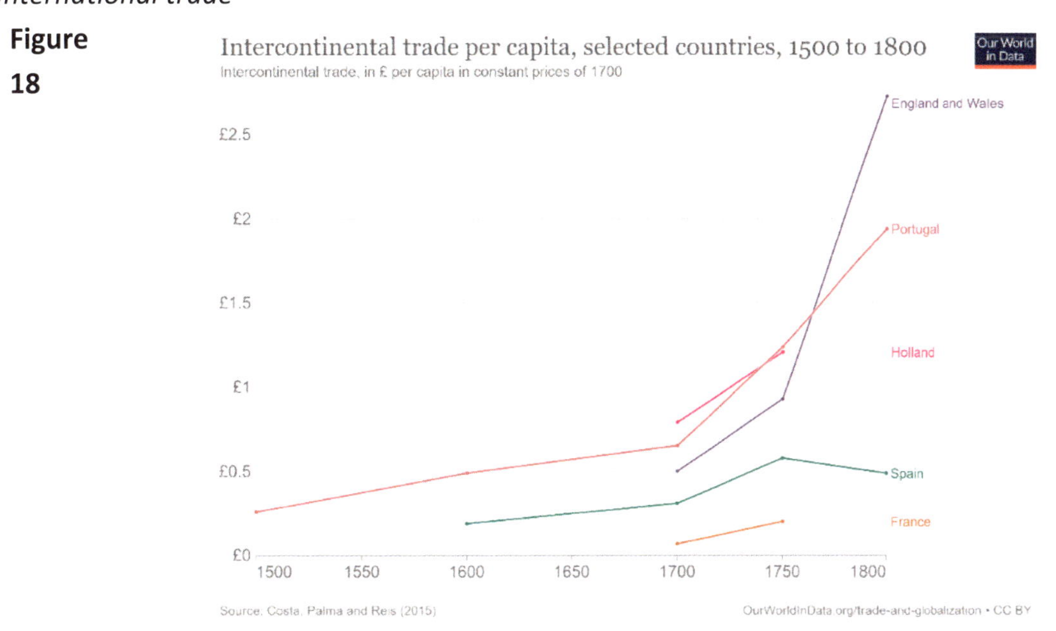

Great Britain dominated international trade. In the 50 years 1825 to 1875 its exports rose as a percentage of GDP from 9% to 23%.

Figure 19

[129] Farrington, Anthony (2002). *Trading Places: The East India Company and Asia 1600–1834*. British Library. ISBN 9780712347563.
[130] Pyne, William Henry (1904) [1808]. *The Microcosm of London, or London in Miniature*. Vol. 2. London: Methuen. p. 159 .

In the United States Cornelius Vanderbilt was in the vanguard of revolt against monopoly and subsidies[131]. When just 28 years old he won an appeal[132] to the US Supreme Court and obtained a declaration of invalidity of the grant of monopoly of steamboat traffic, Chief Justice Marshall declaring that *"Commerce is undoubtedly traffic but it is something more; it is intercourse"*. In 1854 Vanderbilt successfully opposed a Congress grant of subsidy for E K Collins' transatlantic steamship line. Vanderbilt *"unchained the fetters that held men and commerce down"*.

The following graph is of 'trade openness" being the ratio of world exports and imports to global GDP. The higher the index, the higher the impact of trade transactions on economic activity. Prior to 1800 on a global basis the index never exceeded 10%. During the second half of the 19th century the liberalisation of international trade had more than tripled its growth[133].

Figure 20

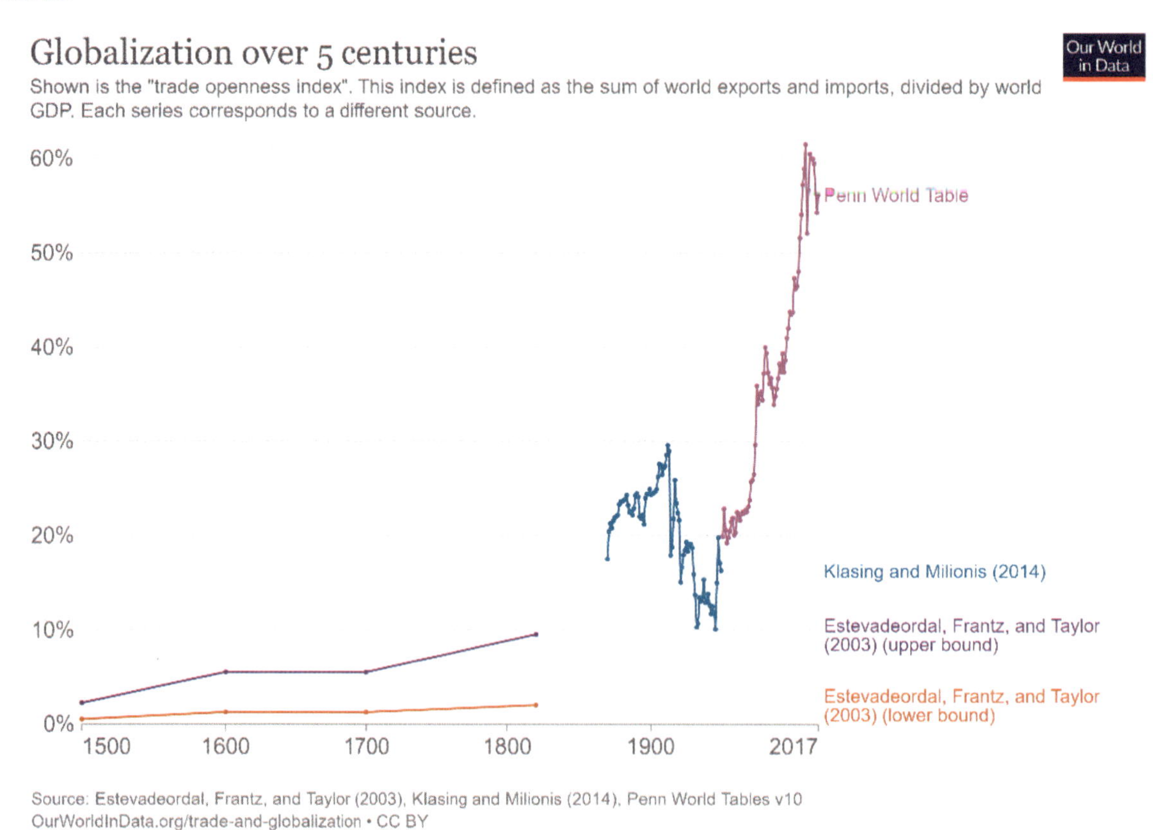

SUMMARY

Mankind's discovery of Nature's secrets heralded the modern age. In less than 50 years from 1800 to 1848 the English industrial revolution brought into existence the open market economy which fostered the means of credit and exchange that sustained it, the liberal democracy that shaped its form and the conditions of free trade that ensured its flowering, all at no cost to the global environment.

---

[131] See Stiles T,J *The First Tycoon* Vintage Books2010 pp 56 – 63 and pp 256 – 264.

[132] In parallel with Thomas Gibbons in 1824 Gibbons v Ogden.

[133] Furceri, Davide; Hannan, Swarnali A; Ostry, Jonathan D; Rose, Andrew K (2021). *"The Macroeconomy After Tariffs"*. The World Bank Economic Review. 36 (2): 361–381. doi:10.1093/wber/lhab016. ISSN 0258-6770. *"Tariff increases are associated with persistent, economically and statistically significant declines in domestic output and productivity, as well as higher unemployment and inequality, real exchange rate appreciation, and insignificant changes to the trade balance."*

PART 4

MAN AND NATURE 1792 - 1988

I The Shock of Change

Conditions of existence

The transformation that overtook England the period 1800 –1840 was both sudden and visible.

In a little more than one generation great masses of agricultural workers, displaced by enclosures of land[134] and the agricultural revolution, moved into the now industrialised towns of the Midlands and the North. The loss of livelihoods with the replacing of hand weaving by machinery not only prompted the Luddite riots[135] but forced the migration into dense conurbations of whole families whose living had for centuries been derived from small industries carried on in their homes and villages.

The sight of the "dark Satanic mills" of belching chimneys and fires seemed to be evidence of a vanishing pastoral life in *"England's green and pleasant land."* [136]

Cottonopolis. Manchester. Impact of the machine age on textile production.

The conditions of existence for the new urban poor in these industrial towns were plainly evident. Uprooted from their rural origins they were subject to privations and distress of a kind never before witnessed. The turbulence and shock of the initial impact of the new industrial age upon

[134] The 18th and 19th century land enclosures transformed communally cultivated open fields into large farms thus permitting more scientific and profitable mixed farming. In addition, much common land over which villagers had customary rights of pasturage or wood or turf cutting, was enclosed. From 1761 to 1796 1,482 Enclosure Acts were enacted by Parliament. In the period 1797 to 1820 there were 1,727. More than four million acres were enclosed under these Acts.

[135] Groups of textile workers who could not compete with the new machine looms and from 1811 to 1816 took to destruction of machinery that deprived them of their livelihood Mill and factory owners took to shooting protesters. It was suppressed with legal and military force, which included execution and penal transportation.

[136] From the poem "And did those feet in ancient time" by William Blake 1804 and printed a few years later.

the working poor were graphically described by the young Friedrich Engels as he observed them when working in in 1842 in his family's Manchester factory.[137]

Passing of a rural age

It is not a coincidence that the period of English great landscape painting opened with the beginning of the industrial revolution. The works of John Constable, Girtin, Cotman and other artists of the Norwich school depict the passing of an age now to be dominated by railways. industrialisation and abandonment of the countryside described in Cobbetts Rural Rides[138]

George Vincent A distant view of Pevensey Bay, (1820). Norfolk Museums Collections

The exodus from rural life into an existence divorced from the outward forms of the natural world not only prompted concern as to its humanitarian consequences but also created a deep sense of unease about the dislocation of the natural order itself. The shock of the New induced anxiety as to what appeared to be the logical and inevitable outcomes of its impact.

The stirrings of concern

Observers could not then detect the extent of the economic effects of the revolution. Nor could they form an informed view about the implications in chemistry and physics of all the processes of industrial activity that were being exploited with each new discovery and invention.

However the social consequences were obvious. The level of population of 6.33m in the year 1300 had been scarcely less than it was in 1700 at 6.76m. Yet by 1790 it had risen to over 12m.

---

[137] In 1842 the 22-year-old Friedrich Engels worked in Ermen and Engels's Victoria Mill, manufacturers of sewing threads. He recorded the privations and distress of the urban poor in the new factories and related housing slums including child labour, despoiled environment, disease and industrial accidents. His book *'The Condition of the Working Class in England'* was the first comprehensive critique of industrialisation. It was published in 1845 in Germany but no translation into English was published until 1891.

[138] William Cobbett campaigned for repeal of the Corn Laws. Moved by the distress of agricultural workers he undertook four journeys 1822 and 1826 through the countryside of Southeast England and the Midlands. He published his account in the Political Register in 1830.

Just a generation later in 1830 it had reached 20m. To those who lived through those times such transformations alone would have caused of as much alarm as of wonder.

Such were the origins of a general concern as to the impact of Man in Nature which later evolved in the last half of the 20th century into what passed as 'environmentalism'. It emerged from the varied responses to the sudden and profound changes brought about by the industrial revolution.

It is possible now to discern in this gradual emergence of concern as to human impact on the natural world some outlines of the then contemporary perceptions of its three principal effects:-

- shock of the explosion of population;
- fear of depletion of natural resources; and
- alarm as to the seeming disturbance of the balance of Nature.

The responses to these perceptions included:-

- the assumption that future outcomes could be deduced from the present;
- the sense that these outcomes were tending to catastrophe; and
- the presumption that only far reaching changes would be adequate to cope with them.

Each of these concerns and responses are examined in this Part 4 as they emerged at the end of the 18th century with the coming of the industrial revolution until 1988.

Study of these concerns and responses reveals the futility of theories about future hypothetical events that rely on predictions, scenarios and constructed models. Not one of the predictions of disaster awaiting humanity as a consequence of the transformation of the conditions of its existence has come about. All have been falsified by events and the passage of time.

Applying the matrix of present conditions to determine future existence has been exposed as a flawed and false process since it necessarily can make no certain prediction of what is as yet unknown. It is futile because it has no utility except as an effective means of promoting alarm and intimidation.

Part 5 below sets out an analysis of how the propagation of global warming ideology and eco-socialism, which took hold in the 4 years from June 1988 to June 1992, came to overwhelm and displace these human sincere concerns as to care for our natural world.

Humanity is surely a form of Nature just as much as the biosphere which sustains it. There is no dichotomy in reality. Man has the capacity for adaptation, evolution, innovation, response to change and survival that are the very characteristics of Nature itself.

That is the evidence of the past 200 years. It accounts for the fulfilment of what did indeed prove to be a true prediction of the future for humanity – that of mankind attaining the broad and sunlit uplands[139]. All this is described in the concluding Part 7 below.

---

[139] *"All Europe may be free and the life of the world may move forward into broad, sunlit uplands".* Winston Spencer Churchill. House of Commons 18.June 1940.

II. Population and Food 1800 - 1974

Malthus

In 1798 Thomas Malthus set out his theory of the consequences of population growth in *An Essay on the Principle of Population*. He held that the famines would become more devastating due to the rate of increase in population particularly among the labouring poor. He asserted that:-

> *"The power of population is so superior to the power of the earth to produce subsistence for Man, that premature death must in some shape or other visit the human race. Population when unchecked, increases in a geometrical ratio. Subsistence increases only in an arithmetic ratio. A slight acquaintance with numbers will show the immensity of the first power in comparison with the second".*

The Malthusian theory of catastrophe is illustrated by the following simple diagram.

Figure 21

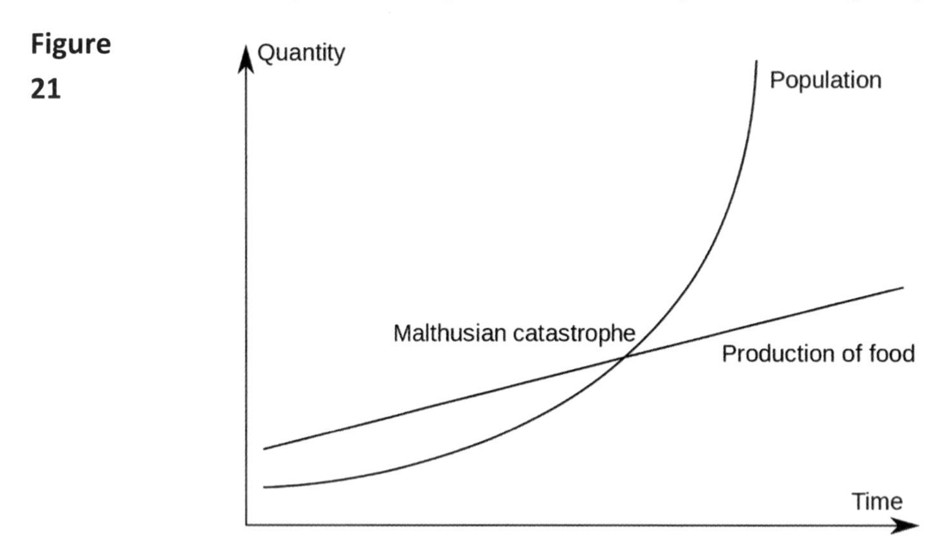

Implicit in the theory was that hunger of the poor would induce more population growth and so would deepen their misery by condemning yet more to hunger.

Such predictions have been falsified by events. The evidence of the succeeding years[140] is that-

- Food supply has increased in greater proportion per person than the growth of population except in regions of dry lands, civil turmoil, war or tyranny.
- Famines have become not more but less frequent.
- Fertility rates have fallen each year since 1962 and are now below replacement levels.

Births and Prosperity

The evidence is that with increasing wealth and improved education countries undergo a demographic transition. Infant mortality declines and fewer births result. Death rates decline as nutrition and health improve. Population grows in these beneficial conditions not *"because people start breeding like rabbits but because they stop dying like flies".* Large numbers of

[140] See Part 7 for a full analysis.

dependent children are not required as a buffer against mortality. Moreover with the advantages of better education and opportunity women marry at a later age and defer having children.

The diagram below shows global population growth rate rising to 2.1% per annum at its height in 1962. The rate of growth then fell to 1.2% in 2010. Between 2018 and 2020 growth rate fell from 1.06% to 0.87%. The current global population (2022) is 7.96bn an 0.83% increase from 2021. It is expected to go into overall decrease within 40 years.

Figure 22

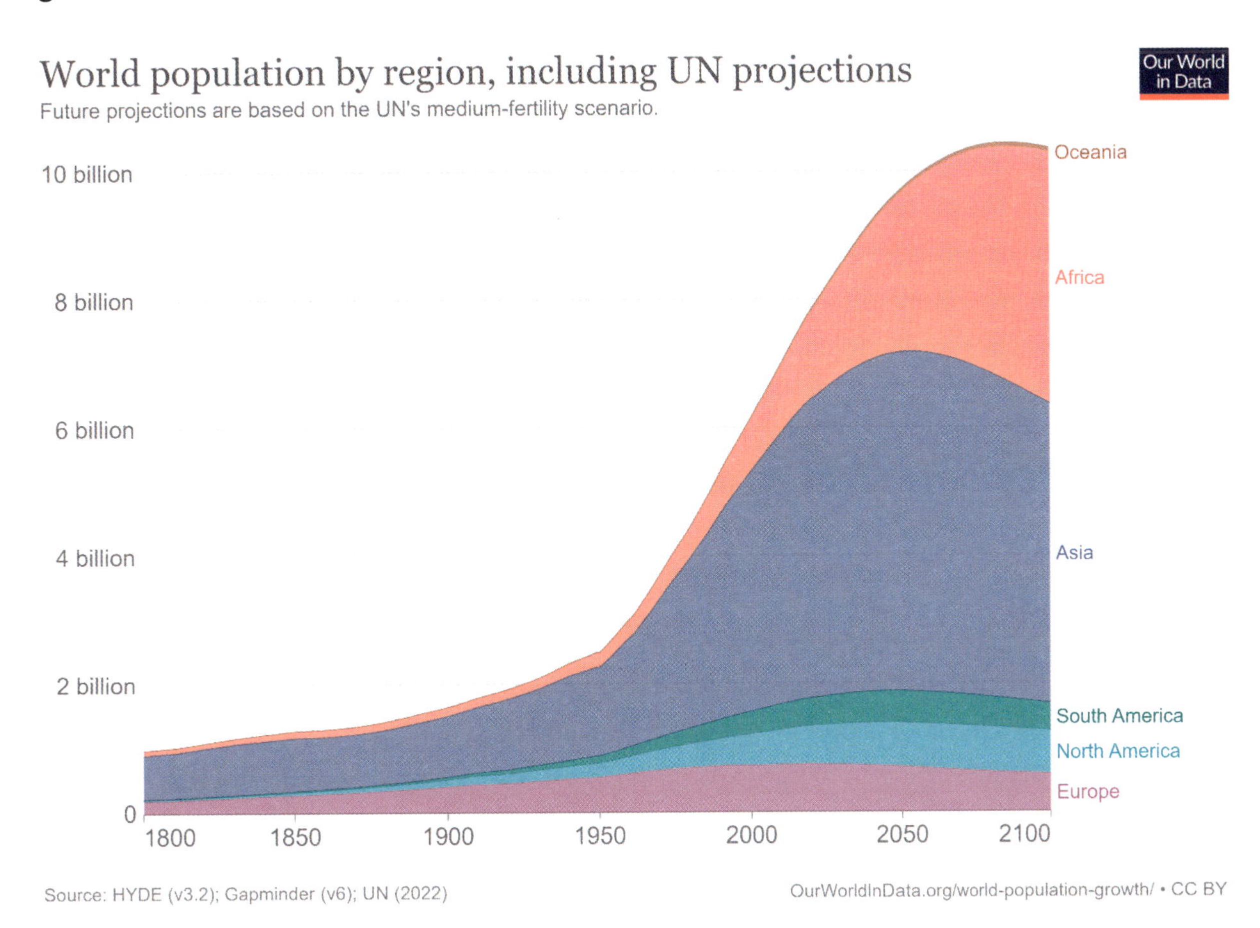

Fertility rates are falling not only in developed regions. In Muslim countries - assumed to be less sensitive to western social and economic influence - rates have fallen by 40%: in Iran by 70% and in seven Arab countries and in Bangladesh by 60%[141]. It is estimated that the global fertility rate will fall below the replacement rate of 2.3 in this current decade[142].

The correlation between decline in infant mortality and average fertility rate in the United States is striking. In 1800 child mortality was 33%: but the fertility rate was 7. By 1900 child mortality was down to 23%: fertility rate 3.9. By 1950 child mortality had declined to 4% and the fertility rate to 3.2. By 2018 child mortality was just 0.6% with fertility at 1.9 - below replacement level.[143]

[141] Eberstadt and Shah *Muslim Population Bust* 2011 Cited in Pinker. S. *Enlightenment Now* 2018 Penguin Random House p126.
[142] Sanyal, Sanjeev (30 October 2011). *"The End of Population Growth"*. Project Syndicate. Gietel-Basten, Stuart; Scherbov, Sergei (December 2, 2019). *"Is half the world's population really below 'replacement-rate*'?". PLOS ONE. https://doi.org/10.1371/journal.pone.0224985
[143] *"Bubble Chart of 'Babies per Woman' vs 'Child Mortality* https://www.gapmonder.org.

Food

Malthus predictions of food supply were confounded by crop yields, effective land use, fertiliser, crop type development and energy. Global population rose from less than 1bn in 1800 to 8bn today. Yet deaths from famine are minute in comparison. Of the 70m people killed by famine in the 20th century 80% were victims of socialist central planning, collectivisation and confiscation.[144]. The British agricultural revolution brought in crop rotation, seed drills and advanced ploughs followed by steam engine mechanisation and opening of markets with the railway booms. By 1920 hydrocarbon fuels were raising yields and driving down costs.

Figure 23 Famine

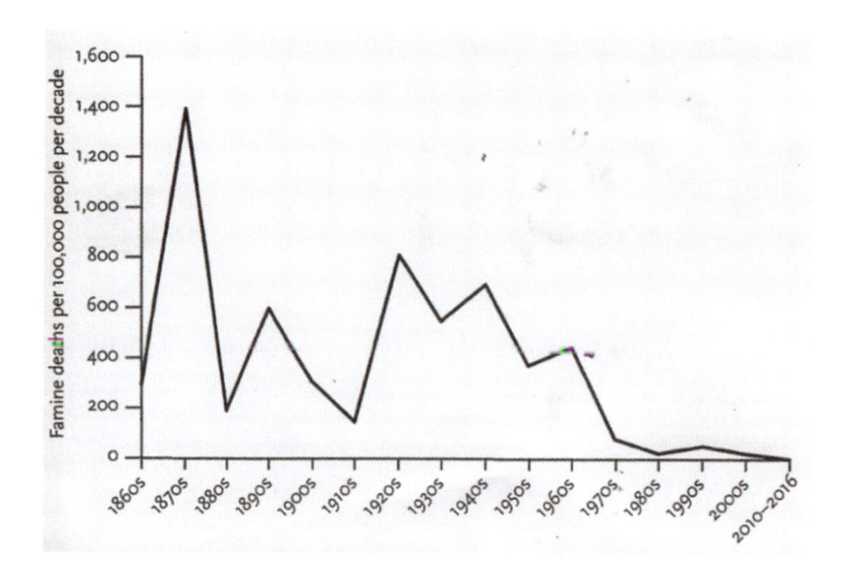

Conversion of nitrogen into fertiliser and modification of crops to produce high yields in the Green Revolution of the mid-20th century[145] including, especially, dwarf wheat and rice, allowing not only vast increases in yields but drastically reduced land use[146].

Figure 24

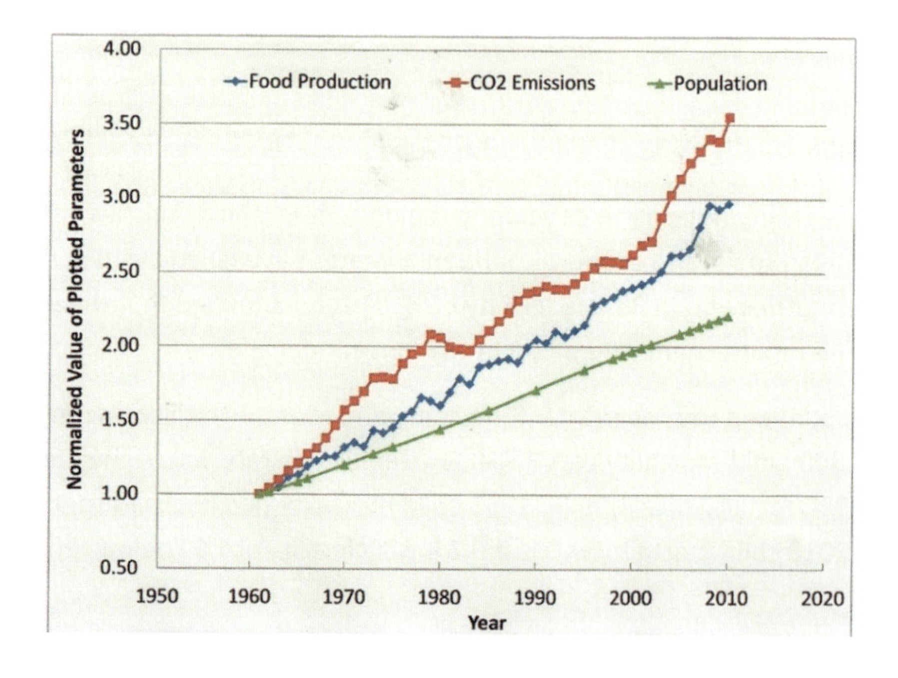

[144] Devereux Stephen. *Famine in the 20th century* UK Institute of Development Studies Working papers 2000 no 105 ISBN 1 85864 292 2.
[145] This miraculous transformation with other advances in human welfare is more fully examined in Part 7.
[146] Center for the Study of Carbon Dioxide and Global Change 2013 www. co₂science.org.

Thus what had taken 25 men a full day to harvest and thresh a ton of grain in 1850 took a modern combine harvester in the United States mid-west a mere 6 minutes[147]. Only one third of the land is now needed that was once used to produce the same given amount of food[148].

In addition, as shown in Figures 42 and 43 Part 7 (page 102) the intensified fertilising effect of increased CO_2 in the last half of the 20th century has dramatically boosted crop yields. This is due both to the enhancement of photosynthesis itself and also to the closing of the apertures of leaf stomata due to availability of more abundant CO_2 so diminishing water loss.

Erlich and the Population Bomb

Predictions of catastrophe driven by rise in population were resurrected in the 1960s.

Stephen Schneider's colleague at Stanford University, Paul Erlich, predicted in 1968[149] that the *"The battle to feed all of humanity is over. In the 1970s hundreds of millions of people will starve to death in spite of any crash programs embarked upon now. At this late date nothing can prevent a substantial increase in the world death rate".*

Included in his predictions was the claim that by the 1980s some 65 million Americans and 4 billion other people would starve to death. Erlich argued – consistent with Malthus – that food aid should not be extended to afflicted poverty stricken regions.

In reality the global death rate continued to decline substantially, from 13 in 1000 in 1965 to 7 in 1000 in 2019. Meanwhile, despite the population of the world having more than doubled, calories consumed per person have increased by 24%.

Ehrlich predicted that India could not possibly feed two hundred million more people by 1980. By 2010 India's population was 1.2bn, three times that of the 400m in 1960, with a total fertility rate in 2008 of 2.6. Rates of malnutrition in India declined from 90% at independence in 1947 to less than 40% in 2010. The absurdity of his predictions has never been acknowledged by Erlich.

It is the unforeseeable that becomes the reality. So it was with the discovery of the means of producing nitrate fertiliser from atmospheric nitrogen, with the Green Revolution which banished famine from India and Pakistan and with the greening of Sub -Saharan desert and drylands due to the increase in atmospheric CO_2.

As has been well said "Genetic engineering can now accomplish in days what traditional farmers accomplished in millennia. Transgenic crops are being developed with high yields, life saving vitamins, tolerance of drought and salinity, resistance to disease pests and spoilage and reduced need for land fertiliser and ploughing. Hundreds of studies, every major health and scientific organisation and more than a hundred Nobel laureates have testified to their safety".[150]

Yet "with their customary indifference to starvation"[151] fanatical eco-socialists have implacably opposed this salvation for millions. They have "starved people, hindered science, hurt the natural environment and denied ...a crucial tool"

---

[147] Deutsch 2011 *People don't need Resources* and other sources cited in Pinker S *Enlightenment Now* Penguin Random House 2018 p75.

[148] Radelet *The Green Revolution Continues* 2015 cited in Pinker S Op cit p76.

[149] Paul R Erlich *The Population Bomb* Sierra Club/Ballantyne Books ISBN 1-56849-587-0.

[150] Pinker S *Enlightenment Now* Penguin Random House 2018 p77.

[151] Ecologist Brand S. *Whole Earth Discipline* New York Penguin 2009 p117 cited by Pinker Op cit.

III. Depletion of natural resources 1865 - 1987

A recurring nightmare for humanity is the spectre of depletion of natural resources deemed to be essential for its continued existence.[152] It re-emerged with great force in the mid 19th century.

Coal

In 1865 William Jevons published his *'The Coal Question'*. He predicted a collapse in prosperity due to *"future exhaustion of cheaply extractable coal"*. In so doing he moved the centre of gravity of fear of the future from alarm as to the rise in population to the scarcity of the energy base of the economic transformation then evident to all. He anticipated the attractive but pernicious and facile notion of 'sustainability' that emerged in the 1980s. He held that it was unwise to allow *"the commerce of this country to rise beyond the point at which we can no longer maintain it"*.

Jevons did not allow that scientific advance would liberate humanity from the restraint of limited resources. He rejected the possibility that human ingenuity, which had brought about the advances he was describing, could also conceive and evolve new forms of energy to replace coal and steam. For Jevons all that could be done was to multiply the efficiency of coal which then would render pointless the discovery or development of any other form of energy resource.

Jevons predicted that Britain would need to produce 102bn tons of coal in the period 1861 – 1970 and that annual production in 1962 would be 2.2bn tons. Actual production in the entire 100 years after the date of his book amounted to slightly less than 2bn tons[153].

Jevons was highly regarded. He has been described as one of the most genuinely original economists who ever lived.[154] Yet he fell into the error of thought that determines future outcomes by a matrix of present conditions. His predictions were being falsified as he wrote. In the early 1830s, Michael Faraday had created the first electric dynamo enabling continuous conversion of rotational mechanical energy into electrical energy. Virtually all electric power is produced using this principle, whatever the prime source of energy - coal, oil, gas, nuclear or hydro. By the 1870s generation of electricity was providing power for manufacturing and for transport by tram and rail.

Growth

The use of logic, based on a premise of existing factors and applied to a conception of the future, is what informed the influential but fallacious *'Limits to Growth'*. It was produced by the Club of Rome in 1972. Contrary to Hayek[155], who contended that only in a responsive open market economy is it possible to adapt and develop in the face of great and complex change, the Club of Rome asserted that the problems facing mankind were of such complexity that traditional institutions and policies were unable to cope with them. It predicted a terminal limit to growth within 100 years resulting in precipitate decline in population and industrial capacity.

---

[152] Marx and Engels did not share the concerns as to the impact of population growth on food supply or the depletion of resources or of the limits of science. Meek *"Marx and Engels on Malthus* 1953 p.82 cited in Darwall R *The Age of Global Warming* 2013 Quartet Books Ltd pp19/20.

[153] Darwall R *The Age of Global Warming* 2013 pp 22/23.

[154] See the comments of Joseph A Schumpeter in *History of Economic Analysis* 1954 p 826.

[155] Hayek F. *Road to Serfdom* 1944 Routledge. See Part 6 below.

The Club of Rome rejected the notion that innovation, invention and advances in science and technology would enable resources to more than meet demand - as the digital age later demonstrated. *Limits to Growth* predicated a post-industrial age in which human development was to be frozen. It prompted a UK government report (1972) that *'a fundamental and painful restructuring of our industrial society is necessary if mankind is to survive"*[156]

As with the Malthus, Jevons and Erlich these forecasts were vitiated by a flawed conception of growth and resources. It is not 'growth' but production and consumption that use 'resources'. But that would be the case with ecological and economic no-growth stability. As a resource become less abundant its price rises. This prompts conservation and also search for less accessible but now economic deposits or less costly and more plentiful substitutes. Moreover it is a fallacy that people need 'resources'. What they need is ideas[157] as to ways of improving their conditions of existence by evolutionary responses to render possible what circumstance dictates to be necessary. 'The human mind, with its recursive combinatorial power, can explore an infinite space of ideas and is not limited by the quantity of any particular kind of stuff in the ground' [158].

Part 7 includes a comprehensive review of the miraculous progress of humanity over the past 4 generations. Food is in surplus in all but dry lands and regions of tyranny, war or civil strife. Energy is becoming dematerialised. Resources far exceed reserves. Production has shifted from the labour intensive to technology and knowledge intensive.

Reserves of natural resources

Projections of shortages implicitly assume resources will only be available at current prices. But the fact of shortages causes prices to rise. At higher prices resources extend to meet demand. As Thomas Sowell points out *"a larger quantity is usually supplied at higher prices than at lower prices, whether what is being sold is oil, apples, lobsters or labour."*

As has been lucidly described[159] if practical conclusions as to the reserves of natural resources are to be reached it is the economic concepts of price, cost and present value that must be considered. Just as prices cause us to share scarce resources and their products at any given time, 'present value' causes us to share those resources over time with future generations – without being aware of it. It is present value that determines how much a natural resource will repay what it costs anyone to discover it at any given time. It is the rising price and present value of the resource that determine the extent of proven reserves not the ultimate theoretical calculation of predicted limits and extents of natural resources. Thus at the end of the 20th century – despite more energy being consumed between 1900 and 1920 than in all the previous recorded history of the human race – known reserves of petroleum were more than ten times larger than in 1950.

---

[156] *'Pollution Nuisance of Nemesis'* A Report on the Control of Pollution 1972 p3 Chair Sir Eric Ashby FRS.

[157] ***Imagination reaches out repeatedly trying to achieve some higher level of understanding until suddenly I find myself momentarily alone before one corner of Nature's pattern of beauty and true majesty is revealed*** Richard Feynman

Imagination is more important than knowledge. Knowledge is limited. Imagination encircles the world. "Alfred Einstein

[158] See Pinker S Op cit pp126/7 for text and examples of fracking to replace coal and oil and liquid crystal to replace red phosphor of cathode ray tubes.

[159] See Thomas Sowell *Basic Economics* 2011 Basic Books Perseus Books Group Ch 12 pp 316 – 326 from which this passage it taken.

Energy underwrites human production and welfare. Discoveries of the past 2 decades have transformed the scale of hydrocarbons known to be lying within the UK's jurisdiction.

Figure 25

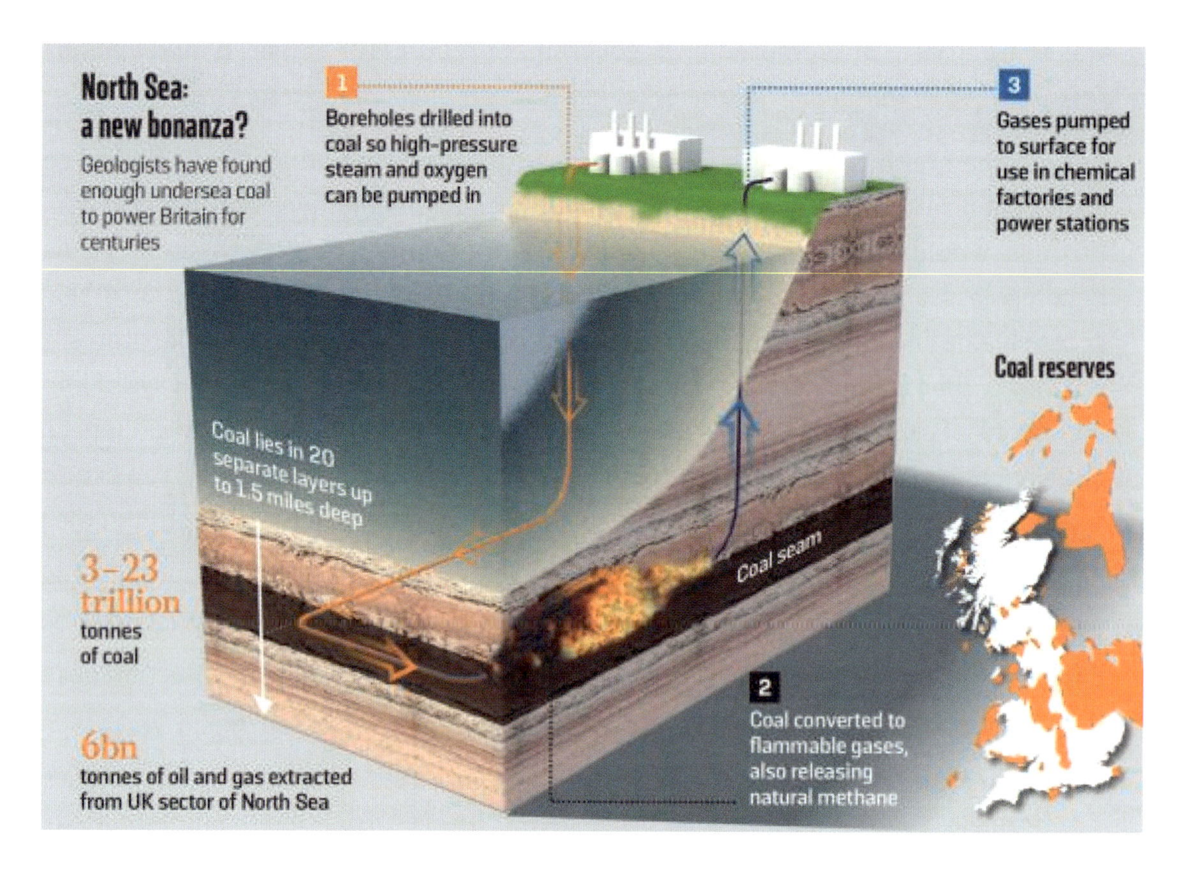

The UK Oil & Gas Authority estimate that just 10% of the gas in the Bowland-Hodder shale in northern England would supply UK gas consumption at present rates for 30 – 40 years. The entire field would, if it were practicable to access it all, be sufficient for over 300 years. Offshore North Sea deposits of up to 23 trillion tonnes of coal have recently been discovered as shown in Figure 25. It is a mercantilism of the intellect – the 'one cake' theory – to suppose there is a limit to the scope and variety of human invention and discovery. No mineral has ever become extinct.

'Sustainability'

There is tapeworm of falsehood deep in the body of "sustainability".

It predicates that current rates of use of a resource must be projected into the future until it is exhausted and that accordingly only what is renewable should be regarded as a resource. The implication is that such can be used indefinitely as it is constantly replenished.

That is not at all the lesson of human society over millennia. This demonstrates the mankind abandons a resource when another one with improved simplicity, utility, extent, cost or availability becomes available.

Nor is anything constantly replenished without cost - save for human ingenuity. There is always a trade-off cost as has been demonstrated in relation to the cost of so-called 'renewable' energy power plants (windmills and sun panels) and electrification of road transport by use of batteries.

The popular spectre of resource exhaustion has played little or no role in the long-run evolution of the energy system. Plenty of wood and hay remained to be exploited when the world shifted to coal. Coal abounded when oil rose. Oil abounds now as methane rises. Advocates of nuclear energy and so-called renewables foolishly point to depletion of oil and natural gas as reasons for their own fuels to win. Oil and natural gas use may peak in coming decades but not because Earth is running out of them. [160].

The report of the Brundtland"[161] Commission set out a so-called 'new global ethic'[162] *'Sustainable development is development that meets the needs of the present without compromising the ability of future generations to meet their own needs'.*

Thomas Sowell[163] has said of "racism" *'it is like ketchup'*. It can be put on practically anything – and demanding evidence of itself makes you a "racist". *"Sustainability"* is environmentalism ketchup. It has no sustainable utility as a term of meaning.

The extent of the resources of the Earth are not known. It is the economics of cost, price and present value that determine the extent to which resources are discovered and are transferred into proven reserves. Neither the needs nor the resources of future generations can be determined. No evidence can be produced of what these will be. Conclusions derived from predictions of the future are not falsifiable and thus are incapable of verification[164]. They should not dictate a global allocation of resources in the present.

'Sustainability' provides for no policy which is not comprehensively secured by the free functioning of markets as regulators of availability, price and utility[165] as described above. Creation of wealth is not governed by resources. Nor is it true that the 'needs of the present' are being met in vast regions of the world. What developing regions require are open markets creating economic prosperity with access to cheap energy in free democracies under rule of law and with the means of credit and investment. That is the lesson of the past 200 years.

The ideology of *'sustainability'* demands restraint of growth in the developed world and re-distribution of global income on an egalitarian basis to the 'developing' world. In 1987 when this seductive but pernicious eco-socialist dogma was first preached by Brundtland, a prominent socialist politician, the digital revolution had not even begun. Nuclear fission was the stuff of fiction. Global average life expectancy has increased from 50 years in 1987 to 73 years. In Africa in 1987 there were over 280,000 cases of poliomyelitis. There are none today. In developing countries levels of undernourishment have fallen since 1987 from 25% to a little over 10%. Extreme poverty ($2.15 at 2017 prices) has fallen from 2 billion to 500 million. Population has increased by 60% from 5 billion to just under 8 billion yet GDP per capita in developing countries increased by 360% ($3,400 to $12,250).

---

[160] Jesse H. Ausubel *Program for the Human Environment*, The Rockefeller University, 1230 York Avenue, New York, NY 10021, USA *Renewable and Nuclear Heresies* 2007 International Journal of Nuclear Governance, Economy and Ecology 1 229 – 243 at p235.

[161] Gro Harlem Brundtland was a socialist politician who became the first woman to be prime minister of Norway. She was appointed by the UN as Chairman of its first director of the World Commission on Environment and Development in 1983.

[162] Speech G H Brundtland to UNEP at Nairobi 8 June 1987.

[163] Professor Thomas Sowell, of American African descent, is America' s greatest thinker and most humane and distinguished citizen.

[164] Popper Karl. *Conjectures and Refutations* Routledge Classics 2002 pp47/8 . The validity of a scientific theory was its capacity to generate predictions that were capable, in principle, of being refuted by empirical evidence.

[165] See more fully Section 6 which compares eco-socialist anti -capitalism in reality with the open market economy in practice.

We know that there is no limit to energy in the subatomic world. Nor is water a 'scarce' resource. Even if used to provide hydrogen and oxygen the likelihood of its exhaustion is on the scale of infinite. Humanity has moved much closer to the de-materialisation of energy. Not only is this true of power generation but also of the digital age. We do more with less.

Brundtland herself was a socialist politician and vice-president of The Socialist International. She had a short career as a doctor before her career as socialist leader of the Labour party in Norway.

'*Sustainability*' is simply a conceptual ideology, a political doctrine formulated by patronising socialist intellectuals in the developed world of the Northern Hemisphere who ignore the astonishing rise of prosperity achieved in vast regions of the former 'developing' world not under the heel of socialism.

'*Sustainability*' has given rise to all of the eco-socialist notions of global governance, wealth re-distribution and rejection of 'capitalism' that marks the final stage of eco-socialist environmentalism.

But its dogma is false. It prophesies scarcity and creates it. It distorts the free expression of human endeavour that characterises freedom and prosperity. It has been justly called a form of 'green imperialism'. The only inhabitant of the future is fear. It is that which constricts enterprise, imagination, invention and prosperity – not dependence on 'vanishing resources'.[166]'

IV. Disturbance of the balance of nature 1962 - 1987

"Who so beset him round with dismal stories do but themselves confound."
John Bunyan[167] Pilgrim's Progress 1678.

Silent Spring 1962

A tide of concern as to the impact of mankind's intervention in the balance of nature began to rise with publication of '*Silent Spring*' in 1962. Conservation as the response to industrialisation became gradually supplanted by a new environmentalism.

It is possible to detect the two stages of this critical change. The first is examined in this Part 4. It took place from 1962 to 1988.

By 1974, just 12 years after Carson's book[168] conservation as the primary response was being gradually subsumed into a form of hatred of mankind for its supposed declaration of war upon Nature. It was accompanied by a succession of global scares and spasms of hysteria conjured up to create profound alarm at the fate that was awaiting humanity.

None of these were justified by empirical science. As is set out in this Part 4 they were all falsified by empirical evidence. Yet despite being false - and being demonstrated to be false - they created a culture of guilt and a predisposition to alarm and fear about human impact on the natural balance.

---

[166] See above p 45 re the economic falsehood of vanishing resources

[167] Imprisoned for 12 years as a Protestant for not accepting Catholic dogma.

[168] *'The Limits of Growth; Mankind at the Turning Point'* 1974 Club of Rome declared Man to be as a cancer of the world.

Then in just 4 years between 1988 to 1992 there became established a very different environmentalism - a form of eco-socialism advocating change in the world order of governance, re-distribution of income, arrest of industrialisation and overthrow of the open market economy. This revolutionary shift is what is examined in Part 5 below.

DDT

Rachel Carson's *Silent Spring* was a sensational success. But its cries of outrage at human assaults on the natural world were not supported by empirical science. Her text was a compound of human guilt and sentimentality as to a lost age of projected beauty and contentment[169]. Her claims were simply bizarre. Carson declared that *"germ cells"* were being shattered by chemicals and that these were *"altering the very material of humanity"*.

It was her assault on the use of pesticides that brought about the banning of DDT. After the introduction of DDT in 1946 in less than a decade control over mosquito carrying malaria in Ceylon (now Sri Lanka) had reduced the death rate by 34% in the first year alone: by 1954 is had fallen by 59% in less than 10 years.[170] When it was banned a potent constraint on the devastating prevalence of mosquito borne malaria in Africa, where it was rampant, was thereby denied to its sufferers.

Carson overreached both reason and evidence in claiming that "One in Four" was the likely incidence of cancer due to release of carcinogens into the environment whilst making no mention of cigarettes. Such 'ideological zealotry'[171] was decisively refuted[172] but it nevertheless gained such a hold on belief as to become a gospel and a cult.

Acid Rain

The theory of acid rain and its propagation was a model for the cult dogma of global warming.

Combustion of coal produces sulphur dioxide which if oxidised into sulphur trioxide (SO_3) is converted into sulphuric acid. Accordingly coal fired power stations use flue-gas desulphurisation and fluidised bed combustion to remove more than 95% of gaseous sulphur from their stack gases.

The conjuring up of the nightmare of falling acid rain due to reckless industrialisation was initiated by a 1971 'report' by the future chairman of the IPCC. The UN Stockholm conference a year later calling for drastic scaling back of fossil fuel combustion. The catalogue of assertions made as to the damage caused by acid rain, all unsupported by scientific observed data, included loss of salmon in rivers and lakes and death of trees and forests.

Yet the actual empirical evidence was that Scandinavian forests had seen growth of 25% due to the nitrogen fertilisation effect of rain. The evidence of soil acidification that was actually being

---

[169] *"To stand at the edge of the sea, to sense the ebb and flow of the tides, to feel the breath of a mist moving over a great salt marsh, to watch the flight of shore birds that have swept up and down the surf lines of the continents for untold thousands of years, to see the running of the old eels and the young shad to the sea, is to have knowledge of things that are as nearly eternal as any earthly life can be."* Rachel Carson 1962 *Silent Spring*

[170] Erlich P. *Population Bomb* 1968 p33.

[171] Darwall R. *Green Tyranny* 2019 Encounter Books p 40.

[172] Doll R et al *The Quantative Causes of Cancer: Estimates of Avoidable Risks in the USA* Journal of the National Cancer Institute Vol 66 No 6 June 1981 pp 1194 – 1308 Darwell R Op cit p 40 "Examination of the trends in American mortality from cancer over the last decade provide no reason to suppose that any major new hazards were introduced in the preceding decades, other than the well recognised hazard of cigarette smoking".

caused by the monoculture impact of pine and fir was rejected. The distinguished Norwegian geologist explaining this was greeted by booing when he addressed a conference on acid rain[173].

The environmentalist consensus on acid rain and human guilt finally collapsed with the publication of a research paper on soil formation.[174] It found that the attributed effect of acid rain was in reality the result of natural soil formation and due to changes in land use.

The research paper found that acid rain with increased deposits of acid and sulphate was *'theoretically unsound and is not supported by direct observation'.* Damage to the lakes and streams had been caused by the process of recovery from "ecological aberration". Normal acidity of the forest floor and soil run-off had returned after tree and stump removal which had reduced forest floor acidity. *'These changes often dwarf in importance the impact of acid rain'.*

The report of the US National Acid Precipitation Assessment Programme in 1989 supported in every material respect the findings of the research paper's authors, Krug and Fink. Yet the US Environment Protection Agency not only suppressed the evidence but also attacked the reputation of Krug such as to compel an apology by the EPA's assistant administrator that its sham 'peer review' was not justified[175].

The acid rain hysteria shared in all essential characteristics, but one, the descent into the irrationality that was global warming cult dogma.

- It used a UN authority to promote an extreme hypothesis as to impending catastrophe.
- It took hold by means of claiming an overwhelming consensus as to 'evidence'.
- It depended on sensationalising natural events and predictions of disaster.
- It was driven forward by the object of elimination of fossil fuel combustion.
- It suppressed dissent and contradictory empirical evidence.

But in one respect if lacked a critical characteristic. It was a propagation of a hypothesis of disaster that was actually happening in the present. It was therefore capable of falsification by empirical observation. It was just this that brought about its collapse.

When, with the Rio summit[176] of 1992, environmentalism took on the armour of global warming dogma its warriors took care that it could not be pierced by empirical contradiction by resting it entirely on fabricated predictions of catastrophe and not on any verifiable evidence.

Nuclear 'Winter'

1976 marked the end of the very cold period of 30 years following the end of World War II. However the widespread panic as to likely descent of an ice age[177] was still fresh in memories.

---

[173] Professor Ivan Rosenqvist 1916 – 1994 cited by Darwall Op cit p68.
[174] Krug E. and Fink C. *Acid Rain on Acid Soil: A New Perspective* Science 1983 Aug 5 ;221(4610):520-5.August 1983 Science doi: 10.1126/science.221.4610.520. National Library of Medicine .
[175] See Darwall R Op cit at pp 75 – 82 for a brilliant analysis of this entire episode.
[176] See Part 5.
[177] See below page 51.

Predictions began to be circulated in the early 1980s that any nuclear conflict between the US and Russia would cause thousands of firestorms resulting in smoke and soot blocking solar radiation. This would cause a 99% reduction in the solar radiation reaching the surface of the planet only clearing over several decades.[178] Summer temperature could decline by up to 20 C (36 F) in the US, Europe and China, and by as much as 35 C (63°F) in Russia. Agricultural production in the Northern hemisphere would be virtually eliminated. It would be equivalent to an asteroid striking the Earth which was then believed to have caused the extinction of the dinosaurs.[179]

There was no empirical evidence whatsoever for the 'nuclear winter'. Indeed there was the contradictory evidence of the colossal volcanic explosion of Krakatoa in 1883 which had created emissions of comparable aerosols. The 'nuclear winter' propaganda[180] was intended to thwart the policy of the US government to deploy Pershing and Cruise missiles as defence against Russian SS20 missiles in an essential part of the strategy of ending the Cold War - which indeed succeeded.

The collapse of nuclear winter theory came with exposure of its fallacy in the light of contradictory observed scientific data[181]. Whilst models adopted by nuclear winter obsessives described how black particles in the stratosphere could block sunlight the evidence was that the soot was not an observed consequence of nuclear explosions but simply of the model programming. It was found that *"The assumptions underlying the study virtually guarantee the occurrence of a nuclear winter"*[182]. Even Schneider described the "apocalyptic" nuclear winter conclusions as having a *"vanishingly low level of probability"*[183].

Political objectives had engendered the nuclear winter scare and maintained it by predictions of catastrophe. It was a further manipulation of science for raw political ends.

Imminence of a new Ice Age

The 30 years following World War II saw a steep fall in global temperature. In the UK the winters of 1946/7 and of 1962/3 were the coldest since 1740. There was widespread alarm at the impact of upon crop yields and of runaway freezing causing famines in developing countries.

The 'New Scientist' had predicted in 1975 - just one year before global temperature started to rise - that *"the threat of a new ice age must now stand alongside nuclear war as a likely source of wholesale death and misery for mankind"*[184] - words repeated 13 years later at the Toronto conference to describe the imminent disaster of global warming.

Schneider himself, as a leading 'climate scientist' himself, had already warned in his 1971 paper published in 'Science' that minute particles emitted due to fossil fuel combustion would diminish

[178] Alan Robock. *"Climatic Consequences of Nuclear Conflict"*. climate.envsci.rutgers.edu.
[179] Crutzen P.J. et al *"The Atmosphere after a Nuclear War: Twilight at Noon"* cited by Darwall R Op cit p107.
[180] Promoted at the 1983 Washington conference. Carl Sagan and Paul Erlich were at the forefront of the campaign.
[181] Seitz R. Comment and correspondence *Foreign Affairs* Vol 62 No 4 Spring 1984 pp 998/999.
[182] Singer F.S. *Re-Analysis of the Nuclear Winter Phenomenon* Meteorology and Atmospheric Physics 38. 1988 pp228 – 239.
[183] Schneider S.H et al *Nuclear Winter Re-appraised* Foreign Affairs Vol 64 No.5 Summer 1986 p983.
[184] Booker C. *Global Warming* Op cit p 18 and 34n.

solar radiation reaching the surface of the Earth and result in severe decline in global temperature[185]. This he held was likely to cause a disastrous cooling by as much as 3.5^{0}C.

The coming ice age became a 'consensus' reinforced by Schneider himself in his 1976 book '*The Genesis Strategy*'[186]. He was described as *"reflecting the consensus of the climatological community in his new book"*[187]. Schneider warned that *"The most imminent and far reaching [danger] is the possibility of a food crisis with unprecedented price hikes in food prices [and] famine and political instability for many parts of the non-industrialised world".*

However once the global warming dogma replaced ice age hysteria in the 1980s Schneider became a prominent IPCC 'lead' author and advocate of the new cult. He was coordinating lead author for the IPCC third 'report' and was co-anchor for the fourth 'report' at the time of his death in July 2010. Schneider's entire propaganda of global warming rested on the assertion that the 20th century rise in global heat had been wholly unprecedented.

As has already been noted[188], it seemed to trouble his conscience not at all that just 13 years earlier in a major book on global cooling[189] he had reported that the last 15,000 years had been unusually warm compared to temperatures for the last 150,000 years and that the last 200 years had been unusually warm when compared to the last 1,000 years – including the Medieval Warming period. Schneider had declared that there was considerable evidence that 'this warm period' had now passed[190].

It now suited his ambition to bury these conclusions as he had done with the process of physics known as 'saturation' which of itself precluded global warming.

The Ozone 'hole'

The 1987 Montreal conference was convened in response to grave alarm as to the reported existence of a large 'hole' in the ozone layer in the stratosphere over the Antarctic. Its Protocol provided for the elimination of chlorofluorocarbons and chlorinated hydrochlorofluorocarbon emissions alleged to be dissolving ozone.

The 'hole' was in reality a thinning of the layer of ozone arising due to natural variable conditions. The 'hole' was reported following the advent of satellite sophisticated observational data recording which disclosed it. The causes of ozone depletion and restoration cycles are not governed by human emissions of chlorine compounds. It is more than probable that the phenomenon has been in play for millions of years.

Ozone in the stratosphere is a barrier to solar UV radiation. UVC harmful radiation is blocked by diatomic oxygen or by ozone (triatomic oxygen). The ozone layer also blocks most UVB radiation which causes sunburn and skin cancer. UVA reaches the surface and is hardly affected by ozone.

---

[185] Rasool S.I. and Schneider S '*Atmospheric Carbon Dioxide and Aerosols: Effect of large increases on Global Climate*'. Science 173 138 –141.
[186] Schneider *Genesis Strategy* 1976 Plenum Press, New York 1976 ISBN 978-1-4615-8758-3.
[187] Dr Tim Ball "*Schneider The Person who set the stage for Entire Deception of Human-Caused Global Warming*". Watts up with That April 11 2019 quoting article in New York Times July 18, 1976.
[188] Part 1 pp 6-8.
[189] *The Genesis Strategy* 1976 Op cit.
[190] Deborah Shapley article in New York Times July 18 1976

Ozone is created in the upper atmosphere in a process called photodisassociation. The collision of ultraviolet (UV) radiation with free oxygen molecules (O_2) causes them to split into single oxygen molecules (O), which combine with other O_2 molecules to create ozone (O_3). UV is the major factor in the creation of O_3. Formation of ozone occurs principally between 15 and 30km altitude in the stratosphere. Densities vary horizontally and vertically and levels over any region constantly change with air movement in the upper atmosphere. The key factor is that the ozone layer is self-healing. As UV penetrates the atmosphere it encounters more free oxygen. By 15 km above the surface over 95% of UV has been expended in the creation of ozone.

Further causes of changes in the density of the Antarctic ozone layer are the intense wind patterns and circulations associated with the Antarctic high-pressure known as the Circumpolar Vortex. Ozone density is also affected by the Polar Stratospheric Cloud that forms when gasses, including water vapor, sublimate directly to crystals because of intensely low temperatures of minus 70°C and below and pressures over the South Pole. This is also occurring, though with less intensity, over the increasing extent of Arctic ice.

Global warming cult dogma and ozone 'hole' dogma are each sustained by false assumptions which predetermine both human cause and also human guilt. The ozone dogma requires that humanity should be blamed for using compounds of chlorine even though it occurs naturally - as with CO_2 - and is essential to medical care particularly the safety of drinking water. Moreover emissions of chlorine from natural causes far exceed those caused by human activity.[191]

SUMMARY

Developing nations owe to their open market economies a wellbeing that their inhabitants even in 1950 could not imagine. Population growth has accelerated it. Surplus food provides for those living in such economies and for most who do not. Energy stores are abundant. There were no cancer inducing pesticides. There was no acid rain. There will be no nuclear winter. There was no ozone hole. There was no ice age. There is no global warming and will be none.

[191] *"Omitted from this story of mass destruction is the fact that the amounts of chlorine contained in all the world's CFCs are insignificant compared to the amount of chlorine put into the atmosphere from natural sources"* Solomon S. Crutzen P.J cited in *'The Holes in the Ozone Scare'* 1992 Twentyfirst Century Science ISBN 9780962813405, 0962813400.

PART 5.

EMERGENCE OF ECO-SOCIALISM 1988 – 2022

"Two ways to be fooled are to believe what is not true or to refuse to believe what is true."
Soren Kierkegaard

I. The Static and the Dynamic worlds

William Vogt

The advent of modern environmentalism was announced by the publication in 1948 of the *Road to Survival*[192]. Its author was William Vogt. He was the director of the Conservation Division of the Pan American Union. His was the work which gave to the world the notion of the 'environment' as a conceived entity of natural phenomena affected by human activity. Hitherto it had been a term denoting natural phenomena as affecting humanity.[193]

He described a view of the environment as its 'carrying capacity' in relation to mankind.

It was a conception that was expressed by a formula B - E = C in which B is unfettered theoretical biotic potential of land to produce plants for shelter, clothing and food, E is the environmental resistance to that potential so that C becomes its actual carrying capacity. In this way he re-formulated the Malthus doctrine of limits which had informed the deepening concern as to the rise in population of the 19th century.

At the root of Vogt's perception of Man in Nature was a rejection of the economic structure that had fostered the wealth, opportunity and welfare created with the coming of industrialization. Upon Vogt's foundation was erected the fabric of eco-socialism of the late 20th century. It consisted of the concept of absolute limits[194] applied to estimates of resources and population and the castigation of the free market economy as the engine of destructive growth. Vogt was the prophet of eco-socialism and provided the ideological base for its assault on 'capitalism'.[195]

It is a startling coincidence that in April 1946 Vogt found himself working with a man whose life's work was the very contradiction of the Vogt doctrine of limits and a testament to human qualities of invention, adaptation and spontaneous discovery. Part 7 is devoted to an account of the reality of the golden age in which we now live. None contributed more to the beneficial transformation of the natural world than Norman Borlaug.

Environmentalism debased

There is, however, no doubt that the Vogt view of mankind and of capitalism came to dominate environmentalism from 1988. It was in the 4 years 1988 - 1992 that environmentalism adopted the single issue dogma of dangerous global warming of the planet as a consequence of fossil fuel combustion. It thereby achieved a decisive dominance. It did so by means of 3 transformations.

---

[192] Vogt W. *Road to Survival* 1948 William Sloane Associates New York.
[193] Mann C.C. *The Wizard and the Prophet* Alfred A Knopf 2018 pp 88/89.
[194] The most notable expressions being *The Population Bomb* 1968 and *The Limits to Growth* 1972.
[195] See Part 6 pp76 and 84 for a discussion of the misnomer of "capitalism".

- The public formulation of a scientific ideology as to the effect of changes in CO_2 content of the atmosphere and of apocalyptic predictions based solely on models.
- Creation of an unaccountable global political bureaucracy promoting the ideology and the progressive demolition of the energy base of modern economies.
- Repeated conventions of world governments to renew intimidation by predictions of catastrophe so as to ensure elimination of CO_2 emissions and transfers of wealth.

The critical shift was from sincere and varied concerns as to human activity and the shock of change to the massing of propaganda behind a single eco-socialist dogma of catastrophe. This demanded rejection both of the free market economy and also of the one nation liberal democracy each of which was deemed to have brought it about.

It is instructive to summarise how it was that the eco-socialist dogma of global warming became welded on to the existing notions of environmentalism. An international political force was merged with a fabricated scientific basis. Its authority was vested in an unaccountable entity and rested on a claimed consensus. The eco-socialist revolution requires:-

- an assault on the free market economic base of democratic societies; [196]
- erecting a collectivist global order over the pluralist democratic structures of self-governing nations[197]; and
- vast re-distribution of wealth from Northern advanced economies to the 'developing' world.

Eco-socialism

Eco socialism is an ideology merging tenets of socialism with green politics and ecology. Eco-socialists contend that the expansion of the 'capitalist' system is the cause of social exclusion, poverty, war and environmental degradation. It asserts[198]that the 'capitalist' economy is incompatible with 'sustainability' which requires eco-socialism for it to become a reality.

In October 2007, the International Ecosocialist Network was founded in Paris. The Network's aims were summarised by it in a formal declaration.[199] *"The ecosocialist movement aims to stop and to reverse the disastrous process of global warming in particular and of capitalist ecocide in general, and to construct a radical and practical alternative to the capitalist system".* This may appear so extreme as to be scarcely credible. However it is essentially the same eco-socialist call

---

[196] For a recent formulation see *Climate Change and Capitalism'* Socialist Review 2019 Dr Stephen Mair, Research Fellow University of Surrey - Centre for the' Understanding of Sustainable Prosperity'. *"If we want to stand a chance of building socialism in the near future, we must become eco-socialists and stop catastrophic climate change now. At the same time, to stop catastrophic climate change, environmentalists must also become eco-socialists. The dynamics that drive climate change are core to capitalism. Serious action on climate change will necessarily amount to the first steps of a programme to end capitalism".*

[197] For a contemporary demolition of this tenet of socialism see Dr Rainer Zitelman report Adam Smith Institute 20.April 2022. The Heritage Foundation found that the countries with the highest level of economic freedom had the highest scores of Yale University's Economic Performance Index averaging 66.8. The 'mostly unfree' and 'repressed' countries registered by far the worst at 37.5 and 36.6 respectively. Heritage Foundation, founded in 1973, is one of the world's most influential think tanks and has had significant influence in U.S. public policy making.

[198] See Kovel, J.; Löwy, M. (2001). *An ecosocialist manifesto*. Paris. Kovel, J. (2007). *The Enemy of Nature: The End of Capitalism or the End of the World?*. New York, NY: Zed Books Ltd. ISBN 978-1-84277-871-5. Wikipedia.

[199] *The Belem Ecosocialist Declaration* issued by Ecosocialist International Network December 23, 2008

for de-industrialisation, wealth distribution and overthrow of the open market economy as expressed by Maurice Strong, Schellnhuber and Edenhofer[200] as discussed below.

The eco-socialist ideology asserts that 'capitalism' be replaced by eco-socialism In order that economic priority is given to the fulfilment of human needs, while "staying within ecological limits for 'sustainable' development". It is a catechism for an egalitarian economic, political and social structure claiming to merge human society with non-human ecology as the only sufficient solution to an assumed ecological crisis. Part of its doctrine is the common ownership of the means of production by 'freely associated' producers, and the restoration of the 'commons'[201].

Its notions permeate the texts and statements of two contemporary extreme environmentalists Schellnhuber and Edenhofer[202] more fully examined in Section 1V (p66 et seq). Each of them has had direct influence on the German government[203]. They contributed to the drafting of the Papal encyclical *Laudate Si* issued in June 2015 prior to the Paris CoP which it was intended to influence. The document speaks of 'things reaching a breaking point due to the rapid pace of change and degradation', advocates governance for the whole range of 'global commons' and contends there is an urgent need of a true world political authority. It includes CO_2 in a description of 'highly polluting gasses' and falsely asserts extraordinary rises in sea level, acidification of the oceans, compromising of food chains and melting of polar ice caps none of which have happened nor could conceivably have been caused by rise in atmospheric CO_2.

The Fixed and the Evolving

Despite their antiquity the notions of Aristotle as to the static nature of humanity and the natural world continue to dominate social theory and socialist thought.[204]

Having no comprehension of the highly developed trading Mediterranean connections of his time and their ordering in a competitive market dependent on credit, price and profit, he had no conception of an economic system arising inherently in a process of evolution, adaptation and growth. He assumed that mankind and nature had always existed in their present form so that only the known needs of an existing population should govern economic effort. As Hayek puts it such a view *"lacked any perception of two crucial aspects of the formation of any complex structure; namely evolution and self-formation of order"*.

It was a fundamental tenet of Aristotle that order in the formation and regulation of human interactions was the result of *taxis* – the deliberate organisation of action of an ordering mind. For Aristotle that order could not permit population to exceed determined limits. He had no understanding of the experimental process of innovation or of adaptation to unforeseen change which govern proliferation of wealth.

---

[200]Edenhofer O. Statement to Neue Zürcher Zeitung November 14 2010. *"But one must clearly say that we redistribute de facto the world's wealth by climate policy. One has to free oneself from the illusion that international climate policy is environmental policy. This has almost nothing to do with environmental policy any more"*. See below pp 66,67 and 69.

[201] Commons is an eco-socialist generic description of the resources which required to be accessible to all members of a society including natural materials such as air, water, and a habitable Earth. These are held in common even when owned privately or publicly. Commons can also be defined as a social practice of governing a resource not by state or market but by a community of users that self-governs the resource through institutions that it creates.[3]

[202] See pp 66 – 67 and 69.

[203] Schellnhuber through the German Advisory Council on Global Change (WBGU). Edenhofer as adviser to Angela Markel.

[204] F.A. Hayek The Fatal Conceit 1988 University of Chicago pp 45 - 47

Aristotle's formulation conflicted with the reality of growth and flowering of prosperity in Greece and its outposts. The dichotomy that existed between the convictions of Vogt and Borlaug also existed in the cradle of western civilization. It was that between a fixed immutable view of the natural order and the reality of economic evolution through the process of innovative development, experiment and adaptation to meet unforeseen change.

The fallacies of the concept of fixed 'resources', extrapolations of population and the predictions that they have spawned, underlie the entire concept of eco-socialist 'sustainability' and the prophecies of catastrophe that is to befall humanity for multiplying its species[205].

II. Rise of Eco Socialist anti-Capitalism 1972 - 1988

Vogt's *Road to Survival,* with its demand for demolition of what had been regarded as the very pillars of freedom and prosperity, emerged 4 years after Friedrich Hayek's seminal book *Road to Serfdom* of 1944 in which he set out in compelling terms the case for freedom both in economic intercourse and in political governance. It was Vogt who in 1948, in the first formulation of modern environmentalism, expressly rejected the 'capitalist' economy in flat contradiction of Hayek. Both books were runaway best selling works.

Limits to Growth

Vogt's repudiation of the basis of 'capitalist' economic development emerges again with the publication of The Club of Rome 'report' *Limits to Growth* (1972) discussed in Part 4 above. The 'report' advocated *"a totally new form of human society"*[206]. The dynamic nature of the free open market economy was to be replaced by a conception of a static society.

It had been preceded by a special edition of the *Ecologist* in January 1972 later published in book form selling 750,000 copies. It consisted of a paper entitled *'A Blueprint for Survival'* and was the work of 37 notable experts including fellows of the Royal Society and holders of science appointments at British universities. It argued that a radically restructured society was essential in order to prevent what the authors referred to as *"the breakdown of society and the irreversible disruption of the life-support systems on this planet".*

Simultaneous with *Limits to Growth* was the publication of a report *'Pollution, Nuisance or Nemesis?'*. It was the result of a study commissioned by the UK government and presided over by Sir Eric Ashby FRS in the context of the then forthcoming 1972 Stockholm UN Environment Conference. It has been described as having *'some of the most alarmist language presented to a British government in peacetime'*[207]. It asserted that *'a fundamental and painful restructuring of our industrial society is necessary if mankind is to survive'*.

---

[205] Erlich P. *The Population Bomb* 1968 Introduction *"The battle to feed humanity is over. In the 1970s the world will undergo famine. Hundreds of millions of people are going to starve to death in spite of any crash programs embarked upon now"*.
[206] Meadows D.H. et al 'Limits to Growth' 1972 Pan Books p183.
[207] See Darwall R. *The Age of Global Warming* 2013 Quartet Books p.65.

Second Report of Club of Rome 1974

The first 'Report' to the Club of Rome (Limits to Growth) was supplemented 2 years later by a second 'Report'[208]. *'The Limits of Growth; Mankind at the Turning Point'* called for the prompt re-structuring of human society by creation of a supposed horizontal social system to replace existing vertical hierarchies and for a single global economic system[209]. Its significance lies in the fact that it publicly endorsed the notion of the guilt of mankind. casting it as a malignant tumour[210]. The creation of Man as an 'enemy' was perceived to be essential to global warming propaganda[211] gaining a hold both on opinion and on policy.

Herbert Marcuse and the New Left

Adding an intellectual affirmation to the socialist implications of the emerging environmentalism were the utterances of Herbert Marcuse, a product of the Frankfurt School of critical theory, who gained notoriety as a prophet of the New Left teaching political theory. He had witnessed the failure of Marxism as an applied explanation of society and sought to substitute a new paradigm of socialism which took account of the natural world.

He lent to environmentalism the socialistic condemnation of capitalism as being a war against nature declaring in 1972 at a conference on ecology and revolution that *'violation of the Earth is a vital aspect of the counter-revolution'*. It was Marcuse who asserted that *'the genocidal war against people is also ecocide"*.[212]

First Global Revolution 1991.[213] Club of Rome "New World Order"

Intended as an impetus to the so-called 'Earth Summit' to be convened by Maurice Strong on behalf of the UN the following year at Rio de Janeiro the Club of Rome's further 'Report' in 1991 *'First Global Revolution'* was to be a *'blueprint for the 21st century'*[214].

For the first time it called to arms the cult of global warming in support of demands for supranational institutions to enforce a draconian environmentalist world regime – described as a 'new world order'. A global taxation on energy to drive down industrial and agricultural output would reverse economic and population growth. It set out a prospectus for transformation of democratic forms of government and the necessary erosion of sovereignty in a *'a positive move towards the new global system in which nation states will have diminishing significance.'*

Brandt report 1980

The Brandt Report[215] of the Independent Commission for International Developmental Issues was named after its chairman. An updated form of the report was issued in 2001. The report

---

[208] *The First Global Revolution_ A Report by the Council of the Club of Rome* – Alexander King, Bertrand Schneider – - Random House, Inc Pantheon Books (1991).
[209] Mesarovic C. et al *The Limits of Growth; Mankind at the Turning Point* Chapters 4 – 9.
[210] Mesarovic C. et al Op cit *"The World has Cancer and the Cancer is Man"* was quoted in chapter 1 of the first edition .
[211] See also for an examination of this precept Part 2 and Part 6.
[212]See Darwall R. *The Age of Global Warming* 2013 Quartet Books p 59 and Kellner D (editor) *Collected Papers of Herbert Marcuse* Vol 3 The New Left and the 1960s 2005 p 173 cited therein.
[213] Alexander King and Bertrand Schneider *The First Global Revolution* Pantheon Books 1991.
[214] Press conference Washington DC Sept 16 1991 launching *The First Global Revolution_ A Report by the Council of the Club of Rome* – Alexander King, Bertrand Schneider – - Random House, Inc Pantheon Books (1991).
[215] Brandt Commission on International Development proposed transfer of 0.7% of national income rising to 1% by 2000

concluded that a great chasm in standard of living existed along a North-South divide and that therefore there should be massive transfers of resources from developed to developing countries. It reported that countries North of the divide were extremely wealthy due to their successful trade in manufactured goods, whereas the countries South of the divide suffer poverty due to their trade in intermediate goods which yielded low export incomes.

The report advocated global adoption of the welfare state model and a global system of taxation in response to the impacts of 'exploitation' of the natural world.

Figure 26

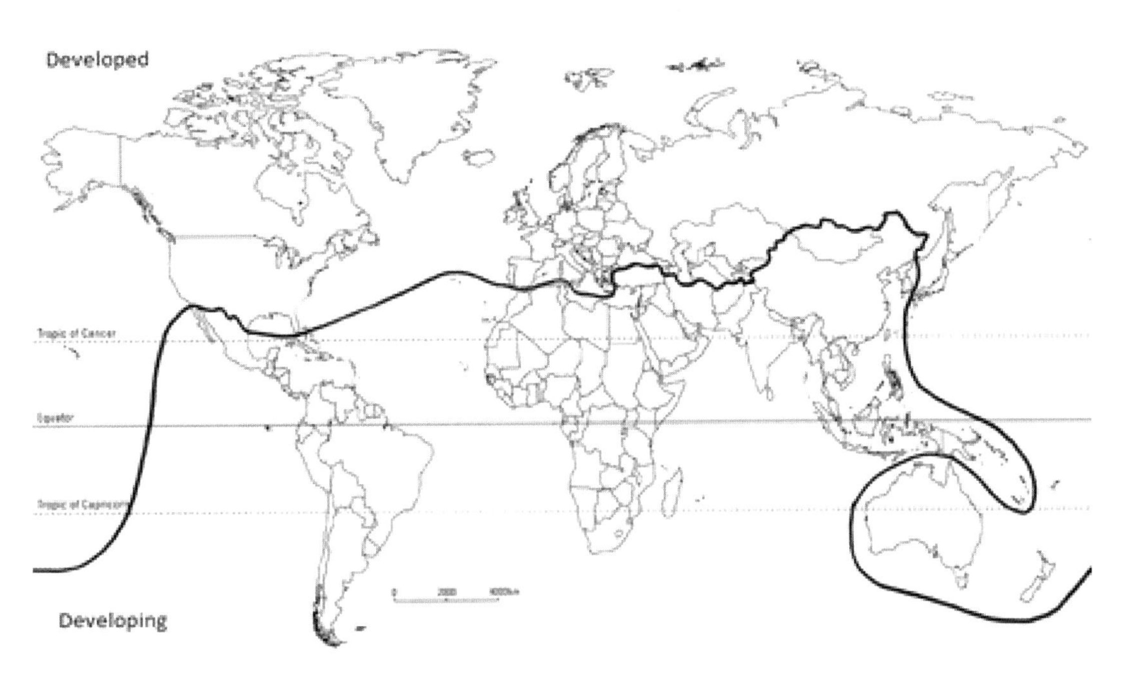

The Brandt Line is a visual depiction of the North-South divide between their economies, based on GDP per capita. It became a model for what became an embedded principle of eco-socialism 12 years later in its final politicised form as it emerged at the 1992 Rio conference.

It remains the implicit basis of absurd demands for 'reparations' for climate change to be paid by industrialised countries in the northern hemisphere which underpin the eco-socialist dogma of Edenhofer and Schellnhuber in the 21st century (see pp 66 – 68 below).

Eco socialism and Maurice Strong

However it was Maurice Strong who ensured that the demise of capitalism, the reversal of industrialisation and superseding of nation state democracy became the ideology of eco socialism embedded in the new dogma of global warming within just 4 years of it first being declared.

In common with the fervid adherents of the dogma at Stanford University[216], in particular Stephen Schneider, as well as James Hansen at the GISS who had worked with Schneider, Maurice Strong was governed by the principle that such outcomes were of such compelling necessity and worth for mankind that they prevailed over the contradictions of science, evidence and truth.

[216] Paul Erlich and Stephen Schneider each in the Department of Biology. See Part 4 above as to their influence.

Strong's success in establishing the dominance of eco-socialism over the existing strands of popular concern about humanity's impact on the natural world was astonishing. Not only did he galvanise 100 world leaders and 20,000 delegates at an international conference convened by him to subscribe to his Climate Change agenda and its eco-socialist prescriptions, he also procured the setting up of the UN climate change bureaucracy with which to bring it about.

He did this without himself having any recognised leading status in that field and within a space of just 4 years. He usurped the disparate strands of environmentalism and replaced them with a single issue belief that came to dominate the policies and culture of democratic western nations to such an extent as to instill complete submission and suffocate dissent.

III. Maurice Strong. Conscription of the UN 1988 - 1992

Maurice Strong was the most significant figure in the creation of the structures and institutions that enabled global warming dogma to gain dominion over rational enquiry and the contradicting laws of atmospheric physics.

Origins and influence

Strong, a Canadian, was born in 1929 during the Great Depression. His family endured severe poverty. His father maintained the family's existence with whatever odd jobs could be found. His mother was a victim of severe mental illness, dying in a mental hospital. His upbringing instilled in Strong a profound conviction of the moral supremacy and justice of socialism. He was from the outset convinced that the scourge of capitalism, as he deduced from the sufferings of the working poor, demanded the intervention of a superior and collective power. In later life Strong declared that growing up during the Depression had radicalised him and that he considered himself to be "*a socialist in ideology, a capitalist in methodology.*"

The vision of the United Nations held out for him the hope and means of creating a new order for humanity[217]. It became his lifelong and paramount conviction. In 1947 he obtained a temporary appointment at the UN New York headquarters. Realising that without qualifications or the advantage of higher education he would have little hope of advancement he returned to Canada and joined as a trainee in a brokerage firm where he developed an informed and intense interest in the oil business.

In just a few years he created out of a small natural gas company one of the largest companies in the industry, Norcen Resources. By 1961 he was president of Power Corporation of Canada, a major investment corporation with interests in the energy and utility sectors becoming its president from 1961 until 1966.

It was then that the Canadian prime minister invited him to take over direction of its overseas development agency so igniting his passionate engagement in the emerging environmentalism of Rachel Carson's *Silent Spring* (1962). In 1971 Strong commissioned a report *Only One Earth: The Care and Maintenance of a Small Planet*, which led to his appointment as Secretary General of the 1972 UN Stockholm conference. Stockholm established environmentalism as an

---

[217] "I knew at once that I wanted to be part of that endeavour" Strong M. *Where on Earth are we going?* 2001 p.156 cited in Darwall R. The Age of Global Warming p 69.

international concern under the aegis of the UN. Strong then launched the UN Environment Programme (UNEP) becoming its first director and a founder member of the Club of Rome.

Strong's UNEP

The UNEP became the entity that produces the IPCC Summary for Policymakers. It is the document that is drafted by a political assembly of representative countries in line with the overriding dogma. Its summary of conclusions are often not drawn from the full text of the 'reports' and in some cases have been shown to be at variance with them. The document is specifically directed to heads of state. It is the principal voice of global warming propaganda.

In 1983 Strong was appointed by the UN to the Brundtland Commission on the Environment. He was determined to bring global warming to world attention. It dominated a conference in 1985 at Villach, Austria sponsored by his UNEP but that conference and its findings were not widely known. Its report claimed a *'likely rise in surface temperatures of up to 4.5^{0}C and rise in sea level of 140cm or more than 4 feet'*. Its conclusion was that it was critical for *'internationally agreed policies for the reduction of the causative gasses'* to be imposed backed by a global convention.

Strong and the UNEP ensured that these recommendations were adopted in the report of the Brundtland Commission in March 1987. The way was open for creation of the eco-socialist scientific ideology, international governance and re-distribution of wealth that Strong welded to environmentalism. His influence at the highest levels of international economic and environmental politics is reflected in the multiplicity of his appointments.[218]

The 'Science' is announced

In 1986, just a few years after the ice age hysteria, James Hansen[219] of the GiSS predicted that the USA would heat by up by a further 2°C - 4°C over a period of 14 years[220]. He claimed that this was due to the greenhouse effect of CO_2 fossil fuel combustion emissions.

Temperature actually increased over that period by just 0.2°C globally (Figure 28) and scarcely at all in the USA[221]. However Hansen achieved notoriety with the sensationalism of his predictions to Congress in June 1988. Over the next 25 years Hansen at the GISS until his retirement and the IPCC used climate models which were cynically manipulated to justify the false hypothesis[222].

All of these modelled predictions are repudiated by satellite evidence.[223]

---

[218] UN Undersecretary, adviser to UN Secretary General, member Advisory Board Harvard's Centre for International Development, senior adviser to president of World Bank, Chairman of the Earth Council, trustee Rockefeller Foundation, board member Davos Economic Forum, member Vatican' Society for Development Justice and Peace, member Club of Rome, Fellow of the Royal Society. Strong was appointed in 1976 as head of the national oil company, Petro-Canada.

[219] Dr James Hansen was director of the Goddard Institute for Space Sciences until 2013. GISS maintains one of the official records of surface thermometer readings. Hanson is an obsessive believer in of global warming dogma. With his colleague, the present director G A Schmid, he has been responsible for the GISS distortions of data described in Part 2 above. He called for CEO's of major fossil fuel companies to be put on trial for *"high crimes against humanity and nature"* for and for spreading disinformation as to global warming. He described coal fired power plants as the *"factories of death"* and the *"trains carrying coal to power stations are death trains"*. *"Climate change is analogous to Lincoln and Slavery – Churchill and Nazism.*

[220] www.realclimatescience.com/2019/02/61-of-noaa-ushcn-adjusted-temperature-data-is-now-fake. and press articles cited.

[221] See Figure11 for 1999 before it was tampered with in 2007.

[222] See Part 2 pp 11 - 13 which examines the tampering with data and graphs by GISS and the IPCC.

[223] See Figure 28 page 68 which sets out satellite measurements of global temperature fluctuation from records maintained by University of Alabama (Dr Roy Spencer).

On 23 June 1988 Hansen gave evidence to the US Senate Committee on Energy and Natural Resources that the 4 hottest years recorded in 100 years had been in the 1980s rising to a peak in 1987. Hansen declared that he was '99% certain' that the cause was the greenhouse effect. The deliberate extravagance of his predictions included his forecast of rise of 2 metres in sea levels by 2000 with hundreds of millions of refugees. Hansen absurd statements – all belied by satellite data - raised intense public alarm and were reported in blanket TV coverage and the press. They succeeded in putting 'global warming' into the public arena from which the IPCC ensured it would not escape. His predictions of catastrophe including his warning to President Obama that *"he had only 4 years to save the world"*. A few days later an international conference in Toronto predicted global temperatures would rise by 4.5^{0}C by 2050.

It had been a mere 10 years since the temperature of the Earth had begun to rise again after the bitter cold of the post war years and the panic as to a coming ice age[224]. 1988 opened the era of 'global warming' - an entirely contrary but equally irrational superstition. It took hold instantly. From 1989 US federal funding of climate change 'researches' rose from $134 million to $2.8 billion in just 3 years – a mere token of what was to follow[225].

UN Governance and the IPCC

It was Strong's declared conviction that establishing global governance was the most important challenge of the next generation. It was for Strong *'one of the great underlying truths of environmental politics'*[226]. Strong's first immense achievement to this end – and one of enduring effect – was procuring that the UN itself should set up the Intergovernmental Panel on Climate Change to propagate the ideology of global warming just a decade after the ice age hysteria.

He did this by exerting pressure through his own creation, the UNEP. It was Strong who ensured that the IPCC was led by a one Bert Bolin the Swedish meteorologist who had issued the false report on acid rain, had convened the Villach conference and who was an obsessed believer of the new dogma.

Strong understood that to prevail it was essential that the cause was vested with the prestige of an international authority. He realised that the authority could not be an entity subject to electoral oversight. He understood that bureaucracy itself was the eco–socialist ideal of global governance.

> *"Bureaucracy is the form of government in which everybody is deprived of political freedom, of the power to act; for the rule by Nobody is not no-rule, and where all are equally powerless we have a tyranny without a tyrant*[227]*."*

The IPCC provided Strong with international sanction of the UN for the spurious 'science' which was to be disseminated over the next 35 years in order to exact undertakings from duped politicians, suffused with climate change dogma, as to 'de-carbonisation' and wealth transfers.

---

[224] See Part 4 p51.

[225] Official US Government figures. See *The Climate Industry: $79 billion so far- Trillions to come* Nova J. Science and Public Policy Institute website - cited in Booker Op cit p 314, 342n.

[226] Strong M. *Our Common Future* 1987 cited in Darwall R. The Age of Global Warming p 69.

[227] Hannah Ahrendt 14th October 1906 – 4 December 1975 was a German-born American political theorist. Many of her books and articles have had a lasting influence on political theory and philosophy. Arendt is widely considered one of the most important political thinkers of the 20th century.

These could not be extracted from totalitarian China[228] and Russia[229] whose scientists knew that the phenomenon of CO_2 'saturation' of Earth's surface radiation precluded global warming dogma entirely and were thus indifferent to IPCC clamour and world outcry knowing that both were groundless and futile.

The IPCC first convened in November 1988. Its promotion claimed that it was composed of over 1,500 of the world's leading 'climate scientists' from all regions of the world who would provide an objective assessment of the evidence on 'global warming'. That had never been the intention and has never been the case.

Its key contributors are drawn from a small pool of academics and government employees who are adherents of the dogma[230]. Most of its 'research' is provided by young associates of little knowledge or experience. Many have been drawn from the UK Met Office and Hadley Centre The IPCC has been cynically used to maintain the illusion of 'consensus'. It is essentially a political bureaucracy issuing Summaries for Policymakers drafted at assemblies of political delegates[231]

No distinguished professor specialising in infra-red spectroscopy at a leading scientific institution who would now stake his career on demonstrating the existence of conclusive evidence of global warming induced by rising CO_2 due to human combustion of fossil fuels.

The propaganda of intimidation

From the outset intimidation was the IPCC's principal weapon. It was a lead IPCC author, Houghton, who stated:[232]*"Unless we announce disasters no one will listen."* Houghton himself drafted the 'Summary for Policymakers' for his first IPCC 'report' of 1990. His models predicted temperature would increase up to 0.5°C every 10 years - a rise every 10 years equivalent to over 70% of the entire temperature increase of the previous 100 years.

With each decade the IPCC apocalyptic predictions of extinction of the human race due to the CO_2 content of the atmosphere[233] become ever more extreme such as to evaporate into fantasy. Yet they have had a fearful impact on the minds of children indoctrinated with the '*deadly impacts of climate change*' [234] in schools and on the uninformed mass of people.

Dogma attains global political dominance

Strong now moved to create the second pillar of the dogma. This was an international treaty committing governments of the world to measures laid before a conference of world leaders. It

---

[228] From Strong who was close to the Chinese government, held a position in Beijing University, had extensive interests in China and retired there.

[229] In rejecting EU political pressure to ratify the Kyoto protocol stipulating reductions in CO_2 emissions Treaty the Russian Academy of Sciences and Alexander Illarianov, Putin's chief economic adviser, declared the "there is no evidence confirming a positive linkage between the lever of carbon dioxide and temperature changes...it is not that which influences the temperature on Earth but it is just the reverse: temperature fluctuations caused by solar emissions influence the concentration of carbon dioxide" International Seminar Moscow 7/8 July 2004 See Booker C. *The Real Global Warming Disaster* 2009 Continuum International Publishing Group at pp 114 – 118.

[230] 24 November 2021 by e-mail in response to the author's enquiries Professor Richard Tol a lead IPCC author explained – concerning the professional or other qualifications (if any) of each IPCC author - that *"The IPCC does not divulge this information. This is partly because the IPCC Bureau is dominated by people from authoritarian countries, and partly because quality of IPCC authors is very uneven".*

[231] See further on this Donna Laframboise YouTube interview Tom Nelson January 2023. Also Booker C *Global Warming* 2017 Ch 14 Op Cit and *The Real Global Warming Disaster* 2008 Op Cit pp-104 – 105, 115-116, 127-127 and 132 - 135.

[232] Sir John Houghton "Global Warming: The Complete Briefing". Lion Press 1994.

[233] See for example the preposterous assertion of catastrophe if a pact is not enforced phasing out coal in OECD countries by 2030 and everywhere else by 2040. "Egypt UN CoP.

[234] See Part 2 p20. See also above footnote re statement of UN Secretary General 7 November 2022 Egypt UN CoP 27 conference.

would also establish a succession of 'Conferences of the Parties' at which the programme of effective de-industrialisation and wealth redistribution could be repeatedly impressed upon governments. At each CoP the IPCC gospel of doom is disguised as 'the science' for the purpose of inducing yet further assaults on CO_2[235]. The latest of these met in Egypt in November 2022.

The ultimate aim for Strong was the replacement of national liberal democracy and its free open market economies with global governance and re-distribution of wealth. He openly declared his conviction quite clearly and without qualification. It is a catechism for eco-socialism.

> *!sn't the only hope for the planet that the industrialised civilisations collapse? Isn't it our responsibility to bring that about?" "[The Earth summit will play an important role] in reforming and strengthening the United Nations as the centrepiece of the emerging system of democratic global governance." "The concept of national sovereignty ....will yield only slowly and reluctantly to the new imperative of global environmental cooperation. It is simply not feasible for sovereignty to be exercised unilaterally by nation states, however powerful."*[236]

China and North Korea

Strong had 'many continuing relationships arising from his career including 40 years of active relationships in China'[237]. He had an office in a Chinese government compound. He was an active Honorary Professor at Beijing University. He retired to China in 2005.[238] He made no secret of his admiration for the Chinese economic and political structure. In 2010, five years before his death, he declared:-

> *We know that pure capitalism hasn't worked. In China, they have used their system – which they call a socialist market economy – quite well to achieve their objectives. It's also in a continuous process of evolution. I've had a working relationship with China nearly all my adult life. They have learned how to use the methods of capitalism to meet their own goals of socialism. China is among the best managed countries today*[239]

Strong was closely connected not only with China but also with North Korea. He had been appointed by UN secretary general as North Korea envoy. In 2004 as president of UN affiliated 'U Peace' Strong established a trust fund dedicated to North Korean projects[240] and hosted a conference in Vevey, Switzerland, on North Korean energy issues.

---

[235] See the latest CoP doom forecast November 7 2022 '*Humanity has a choice: cooperate or perish. It is either a Climate Solidarity Pact – or a Collective Suicide Pact'* UN Secretary General. Speech opening of COP27 Sharm el Sheikh. Echoing William Vogt 1948 *"We have been skidding down the road toward national suicide by destroying the environment that permits out survival.'*

[236] Maurice Strong, Interview 1992, concerning the plot of a book he would like to write. Gibson, Donald. *Environmentalism: ideology and power*. p 95.

[237] https://www.fdd.org/analysis/2007/02/06/at-the-united-nations-the-curious-career-of-maurice-strong/.

[238] He had resigned from his UN posts in 2005 after a web of corruption surrounding the 'Oil for Food' programme came to light. An inquiry conducted by Paul Volcker U.S. Federal Reserve chairman discovered that the Iraq foreign minister had handed $1m in cash to a Korean associate of Strong which was paid to Strong by cheque and endorsed by him in favour of a Strong family company.

[239] Guardian 23 June 2010.

[240] DPRK Trust Fund.

Rio Earth Summit 1992

The so-called Earth Summit held at Rio de Janeiro in June 1992 effected the emergence of the dogma of global warming as an eco-socialist international political force. It was the second of Strong's decisive achievements in raising up of the eco-socialist agenda on to the international political stage.

It was the largest conference ever organised. It was attended by over 100 world leaders and 20,000 official delegates. 179 Countries were represented. Strong was responsible for the entire content, agenda and process of the conference.

He procured that it should set up the UN Framework Convention on Climate Change. The "ultimate objective" of the Convention was to be *"stabilisation of greenhouse gas concentrations in the atmosphere at a level that would prevent dangerous anthropogenic interference with the climate system"*. It promoted limitation of industrialisation in line with 'sustainability' by adopting Agenda 21 and promoting creation of the 'Commission of Sustainable Development'.

The Rio conference transformed environmentalism as an ideal into an international political campaign resting upon a fundamental scientific falsehood. It was driven by the impetus of socialism, maintained by intimidation and gained mastery over empirical scientific evidence.

No longer would international conferences on the environment be forums for discussion and persuasion. They were now exclusively for disseminating eco-socialist propaganda on global warming– a dogma which had secured its dominion in just 4 years from its first public declaration. International conferences were to be convened regularly in order to secure the required undertakings of compliance and monitoring of CO_2 emissions reductions. These CoP gatherings were to serve as platforms for announcements of the fate awaiting humanity if immediate steps were not taken to 'Save the Planet'.

Eco-socialism and the means of its propagation and compliance were established by Strong at the Rio conference.

It constructed the framework for fulfillment of all of Strong's objectives. The agenda was set for re-distribution of wealth and for overthrow of western economies by elimination of their energy base and the empowering of a UN global bureaucracy to ensure its implementation. It was no less than the opening of the campaign by which he intended to bring about what he himself declared to be the essential collapse of "industrialised civilisations". It is the inevitable consequence of implementing Net Zero that such a catastrophe would come about. It can have no other outcome.

Even if the day eventually comes when reality and evidence take dominion again over superstition, belief and dogma, a vast international structure has been created which is maintained not by scientific evidence of global warming - for there is no such evidence – but by the vast outpouring of resources which sustain the colossal fraud and by the innumerable institutions and the hundreds of thousands of those around the world who derive their own wealth and status from its propagation and ensure the suffocation of enquiry, doubt and dissent.

IV. Wealth re- distribution

The basic tenet of eco- socialism

Eco-socialist global warming dogma demands that vast wealth be transferred by developed countries to 'developing' countries. It is a central tenet of eco-socialism. It is merged in the notion of 'reparations' for 'damage' caused by the supposed impact on the climate of the industrial revolution and of industrialisation.

William Vogt had opened the post-war era with forecasts of catastrophe due to population explosion[241] and his bitter assault on wealth and free enterprise[242].

Barbara Ward achieved a global reputation for her promotion of care and concern for the environment. Her report *Only One Earth: The Care and Maintenance of a Small Planet* had been commissioned by Strong for the Stockholm conference. Central to her case was the need on ethical grounds for distribution of wealth created by *'the reckless economic expansion of the last 300 years'* by massive transfers to developing countries.[243]

At the Stockholm conference both Paul Erlich[244] and Barry Commoner[245] – each of them biologists and colleagues of Stephen Schneider – demanded that as capitalism and colonialism had caused population explosion and the environmental crisis it was only right that rich countries should pay reparations for the consequences of these for poor nations.

The recurrent theme of reparations and equitable re-distribution of wealth informs the eco-socialist activity following recovery from the bleak economic period of the 1970s [246]due to the tripling of the price of oil.[247] The conference at Rio (1992) provided that wealth transfer would be on the basis of a split between industrialised and developing countries. Included in these 'developing' countries including China and India - which were and remain largely exempted from the proposals to cut CO_2 emissions.

Essen conference 2009

In June 2009 an eco-socialist conference was held in Essen[248] under the shadow of the financial crisis of 2008. Its theme was the need for seismic changes in western society and its democratic values. The conference, *'The-Great Transformation: Climate Change As-Cultural Change'*, was attended by 450 delegates including members of the Advisory Council on Global Change

---

[241] *We have been skidding down the road toward national suicide by destroying the environment that permits out survival* Vogt W. Road to Survival 1948 William Sloane Asscs p 275.

[242]*Had the parasite of European industrial investment not been able to sink its proboscis deep into new lands world history would be very different*" he demanded that, "*having much wealth*"*...we [the USA] must in human decency use our resources be used to help less well-endowed peoples* Vogt W. Op Cit p 69 and xiv.

[243] Ward B. (Baroness Jackson of Lodsworth) *A New Creation? Reflections on the Environmental Issue* Pontifical Commission for Justice and Peace in Vatican City 1973 pp 16, 31. In 1971 she founded the International Institute for Environment and Development, acting as president from 1973 and chairman from 1980.

[244] Erlich's assault was not just on population but on capitalism *"Actually the problem in the world is that there are too man rich people"* 'Sarasota Herald Tribune *Population Expert Faults Wealthy'* 6 April 1990.

[245] Barry Commoner was among the founders of the modern environmental movement. In his book *The Closing Circle*, Commoner first brought the idea of sustainability to a mass audience. Commoner was a left-wing, eco-socialist , postulating that capitalist technologies were chiefly responsible for environmental degradation. "*The environmental crisis wrenched open the brutality of racial competition for survival. Producing for the common good not for profit would solve it*" Quoted in Darwall R The Age of Global Warming p79.

[246] See for example Brandt Commission on International Development proposed transfer of 0.7% of national income rising to 1% by 2000.

[247] By the end of the OAPEC embargo from October 1973 to March 1974 oil rose by 400%, from US$3 per barrel to $12 per barrel globally;

[248] https://www.americanprogress.org/article/the-great-transformation-climate-change-as-cultural-change.

appointed by the German government. It concluded that climate change policies were not compatible with national democracy.

Its leading speakers proposed the appointment of scientific guardians to be embedded in legislative process. A cultural revolution was required to transform capitalism.

Ottmar Edenhofer,[249] a prominent speaker and chairman of the IPCC Working Group III on climate policy, asserted to the Press that *' one must clearly say that we re-distribute de facto the world's wealth by climate policy. One has to free oneself from the illusion that international climate policy is environmental policy. This has almost nothing to do with environmental policy'*[250]. Comprising broad statements of eco-socialism the conference speeches advocated the essential need for control of global commons[251], fundamental redistribution of wealth and income and the reinvention of society.[252]

In a cameo of what Hayek had predicted[253] as to the fate of liberal democracy Claus Leggerwie[254] expressed a common conviction that the processes of democracy were too slow to permit the drastic change necessary to meet the challenge of climate change. Hans Schellnhuber[255] called for a new constitutional settlement that would enable judges to overturn the will of the majority.

The conclusions of the conference were arrived at using models which had been applied in the then most recent IPCC report (2007). In the graphs set out below (Figure 27) the brown curves represent 'probability distributions' generated by a model, using the IPCC emission scenario, run repeatedly to arrive at range of likely temperatures for 2020 - 2029. The red curves are 'probability' ranges for 2090 to 2099 showing up to 7^0C degrees rise in global temperature.

Figure 27

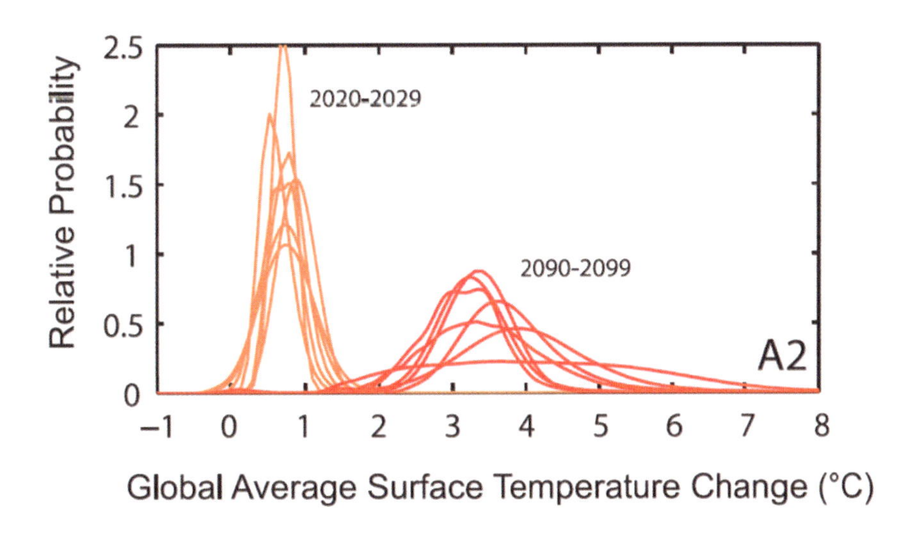

[249] Ottmar Edenhofer an economist and director of the Mercator Research Institute on Global Commons and Climate Change.

[250] Bernhard Pötter *'Climate policy to re-distribute wealth'* Neue Zürcher Zeitung Nov 14 2010 cited by Darwall R. *Green Tyranny* pp19/20.

[251] 'Resource domains' regarded as the common inheritance of humanity including the atmosphere.

[252] Speech of Thomas Homer-Dixson 8 June 2009 https://homerdixon.com/the-great-transformation-climate-change-as-cultural-change/.

[253] See more fully on Hayek in Part 6.

[254] Author of *The End of the World as We Know It: Climate, the Future and the Prospects for Democracy*. Claus Leggewie is a political scientist. Since 2015 has been Ludwig Börne Professor at the Justus Liebig University in Gießen. Until 2017 he was director of the Kulturwissenschaftliches Institut (KWI), Institute for Advanced Study in the Humanities in Essen (Germany).

[255] Senior adviser to the German Chancellor Angela Merkel on climate. Scientific adviser to the European Commission and member of the Club of Rome.

The IPCC model reveals a rise from 2007 of 1.8^{0}C – 2.5^{0}C by 2020 - 2029. However the satellite record (Figure 28) for the entire 43 years to 2023 reveals an actual rise of just **0.33^{0}C**, with fluctuations consistent with solar activity and oceanic shifts including La Nina and El Nino.

Figure 28

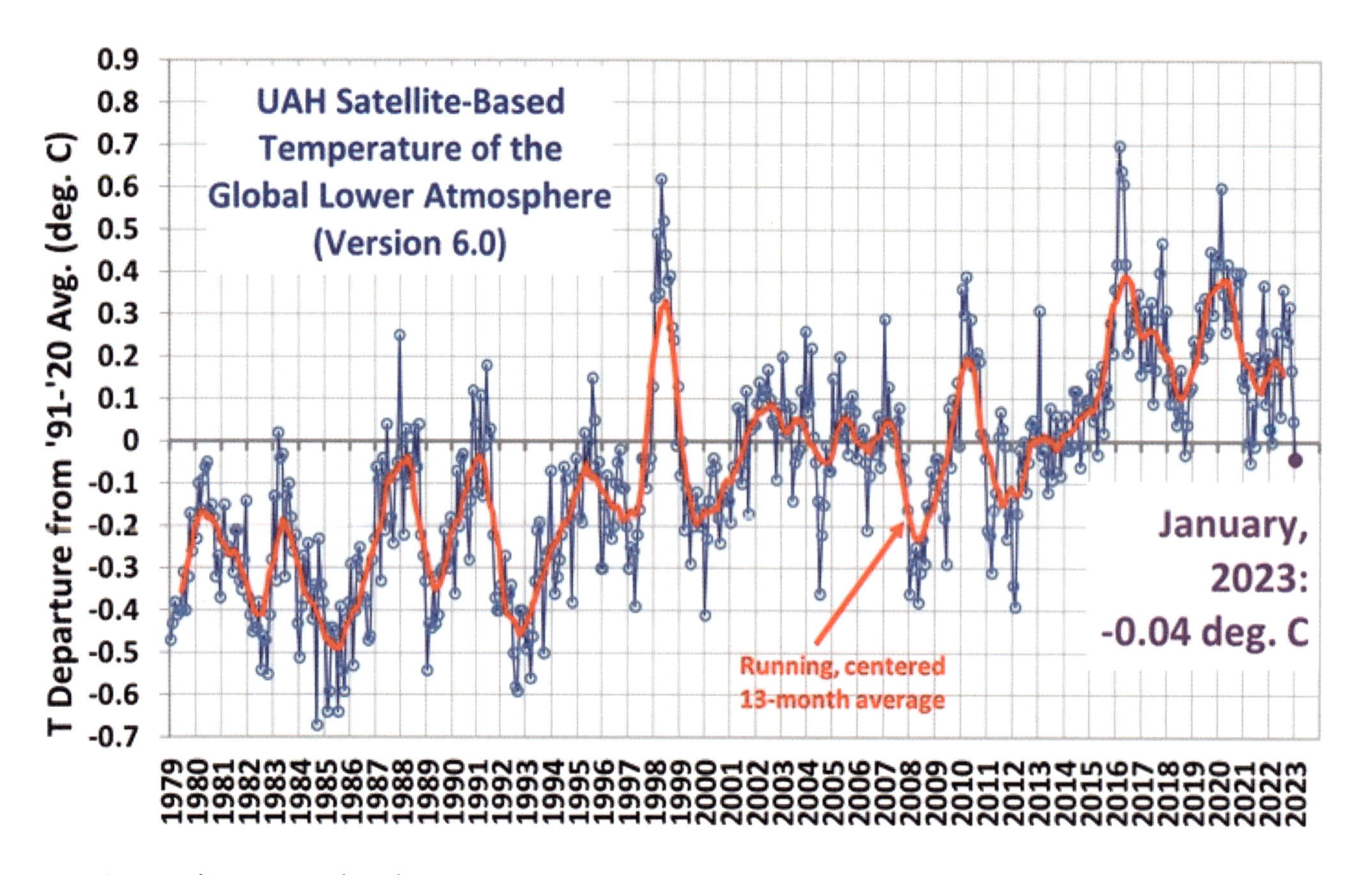

Green Climate Fund and Paris 2015

In 2010 a Green Climate Fund was established within the framework of the UNFCCC[256] to provide a reservoir of funds for draw down by 'developing' countries and funded by developed countries which were parties to the Rio Convention.

A year later the German government's Advisory Council on Climate Change declared that a worldwide reconstruction of economy and of society was required comparable to the emergence of the adoption of the Neolithic agricultural way of life and the Industrial Revolution. It asserted that global governance structures were the indispensable driving force of the transformation of politics economy and society.[257]

In February 2015 referring to the Paris CoP later that year the chief UNFCCC organiser commented:

> *'This is the first time in the history of mankind that we are setting ourselves the task of intentionally, within a defined period of time, to change the economic development model that has been reigning for at least 150 years since the industrial revolution.[258]'*

[256] United Nations Framework Convention on Climate Change established by Maurice Strong and adopted at the Rio conference 1992..

[257] WGBU (initials in German for Advisory Council on Climate Change) *World in Transition: A Social Contract for Sustainability: Summary for Policy-Makers* 2011 cited by Darwall R. *Green Tyranny* p21/22

[258] Statement of Christiana Figueres, Brussels February 2015.

The Paris CoP determined that, for the benefit of 'developing' countries, industrialised countries making the greatest cuts of emissions would also contribute together an annual amount to this fund of $100,000,000,000. Included among the beneficiaries is China. During the following year it commissioned more than one large coal plant every week[259]. In January 2022 it was operating 3,037 of them[260]. China has 69% of global new coal fired capacity – 1,011,037 MW.

Papal Encyclical Laudato Si 2015

The papal encyclical *Laudato si'* was issued in 2015 by the Jesuit head of the Christian catholic church. As has been noted above extreme eco-socialist statements included in the circular were contributed by members of a dedicated German eco-socialist group linked to the Mercator Institute and the Potsdam Institute for Climate Impact Research including the Jesuit economist Ottmar Edenhofer and Hans Schellnhuber. It condemns humanity for causing dangerous change to the climate. It claims that mankind will cause *'unprecedented destruction'* of the natural environment. It advocates governance for the whole range of 'global commons' and contends there is an urgent need of a *'true world political authority'*. It includes CO_2 in a description of *'highly polluting gasses'*[261].

Part 7 below demonstrates the falsehood of these dismal statements.

Eco-Socialism 2022

In an article written in July 2022[262] for Monthly Review the Minister of Consumer Affairs in the Spanish government and leader of the United Left Party of Spain set out again the eco-socialist mantra of the impacts of climate change, of population and of resources scarcity as justification for coercive socialist controls and wealth re-distribution.

Its repeated catechism is that capitalist production exerts pressure on the environment through the emission of CO_2 from burning fossil fuels and by climate change resulting from build-up of greenhouse gases in the atmosphere. Capitalism is causing an *'eco-social crisis'*. This is what has spawned the WEF[263] so-called "stakeholder capitalism" and demands for control of 'global commons' including the atmosphere.

Eco-socialism requires that *'production and distribution processes be socially just'*. Democracy cannot survive if it does not enforce a' *complete program of positive safeguards'* and provide effective solutions to the eco-social crisis. Capitalism has created planetary crisis and this demands eco-socialist governance. The dependency on fossil capital is said to be *'an expression of the fragility of the whole social system'*. In a world with finite natural resources and fossil fuels reaching or exceeding their respective peaks *'humankind is running out of time'* under capitalism. It asserts that capitalism *'deliberately ignores the physical context of which it is necessarily part'*

---

[259] San Francisco-based think tank Global Energy Monitor (GEM) and the independent organisation Centre for Research on Energy and Clean Air 4th February 2021.
[260] https://www.statista.com/statistics/1268457/coal-power-plants-in-china-by-province. Published by Statista Research Department, Jan 31, 2023
[261] See pages 17,18, 19,84 and 85 of the encyclical..
[262] 2022, Volume 74, Number 03 (July-August 2022)._Monthly Review is described as an Independent Socialist Magazine.
[263] World Economic Forum . 'Stakeholders' include the 'global community' another expression of socialism justifying intervention.

citing ozone layer holes, ocean acidification, aerosols and climate change though none of these has any scientific validity[264].

A recent description[265] of what eco-socialism requires concerns priorities in the UK. Public-investment banks and trade unions would ensure the evolution of national economies in the direction of environmental sustainability. Calls for environmental laws by demonstrators are not enough. *'The transition needed is a day-to-day process requiring public control over the allocation of spending and the content of firms' decisions'.* Typical are the demands for a new social system and the overthrow of the free market economy as the response to climate change and the *'global devastation that the accumulation of capital causes'.*[266]

Socialism has found a new 'inherent contradiction' for 'capitalism'.

It is as false as it was for Malthus, Jevons, Vogt, Erlich, the Club of Rome, Brandt and Brundtland. Energy is being dematerialised. Food yields are increasingly abundant. Rising rates of population growth ceased in 1962. Labour intensive production evolves to technology and knowledge intensive production. We produce more from less. Billions of humans have risen to prosperity. Social conditions have been transformed. Poverty, hunger and disease are in steep decline.

The paramount factor is Nature itself. Nature has the supreme capacity to adapt, mould, evolve, multiply and survive. That is the evidence of all the ages. It is to be found in all its creatures and none more so than in the species of *homo sapiens.*

Conclusion

Eco-socialism betrays sincere human concerns about Nature. Preaching hatred of mankind it has opened deep fissures in social unity. By rejecting what has raised up humanity it has fractured pride and stifled gratitude for those who brought this about. By rendering others guilty it has assumed a hideous mask of virtue. The cult of climate change demands a denial of all dissent. Free expression of contradictory evidence is stifled and even condemned in public discussion, the Press and broadcasting. The scientific method has been debased to *'foregone conclusions of a faith: conclusions that are not to be questioned but only surrounded with fictitious scholarship and secured against disproof'*[267].

We are deaf to the warnings of Hayek. We are careless with what we should most treasure.

SUMMARY

The dominion of eco-socialist global warming dogma over science has subverted sincere human concerns induced by the shock of the profound transformation of human existence. It is a cult belief founded on dogma not evidence. Eco-socialism seeks to replace one nation democracy with a form of global governance, to subvert the open market economy and procure vast wealth re-distribution. It admits no contradiction or doubt. It is based on a colossal falsehood.

---

[264] See Part 4 pp 48 – 53 above. See also e.g Greenpeace founder Patrick Moore *Fake Invisible Catastrophes* 2021 Ecosense Environmental pp172 – 197. Also Appendix to Nieboer J. *CLIMATE All is Well All will be Well* 2021 Bruges Group.

[265] *"Ecosocialism more than another third way. Preparing for government"*. Social Europe. Christophe Sente and Timothée Duverger 27th September 2021 socialeurope.eu/ecosocialism-more-than-another-third-way preparing for government.

[266] See e.g Professor Kolin A. *Irrationality of Capitalism and Climate Change: Prospects for an Alternative Future* –Lexington Books 15 Feb 2022

[267] Scruton R. The Uses of Pessimism 2010 p170

Part 6.

ECO- SOCIALISM & THE OPEN MARKET ECONOMY

"Socialism exalts the rule. Liberalism exalts the man."
Winston Churchill 4 May 1908 speech in Dundee

I. Opportunity and Insecurity

The progress of civilisation in just a few generations is beyond our capacity to marvel at or express. It ranks in wonder with the emergence of a butterfly from a chrysalis.

The Industrial Revolution, the coming of the open market economy, adoption of free trade and the rise of liberal national democracy all occurred in England within a span of 50 years. That is 0.025% of the entire span of human existence. As these advances took hold in other countries and regions the conditions of existence of humanity, which had prevailed for 600 generations since farming settlements first were formed, were utterly displaced in just 3 or 4 generations.

It can hardly be wondered that such profound, sudden and irreversible change should have prompted forebodings of catastrophe and notions of guilt to grow like fungi on the insecurity, incomprehension and fear induced by the shock of its impact.

Being irrational these responses are not susceptible to reason or evidence.

Dichotomy of responses

The sudden emergence of the modern world over the globe in the last 200 years has induced a profound dichotomy of responses both to the abundance of opportunity it ushered in and also to the sense of insecurity and disorientation engendered by its disturbing changes.

The predominant response of the inheritors of the Industrial Revolution at first in England and then in Germany, France and the United States was to embrace the opportunities that it afforded to create the greatest advance ever made by mankind to a more beneficial existence.

However, shocked by dislocation of the known world and by fear of the unknown, humanity also sought security through resisting and arresting the forces of change while condemning those deemed guilty of its onslaught. Being confronted with a world in which the way of life and conditions of existence of thousands of years were being suddenly and for ever overturned gave rise to a sense of deep insecurity.

The eco-socialist prophets have subverted these genuine anxieties into a dogma of impending climate disaster requiring both a reversal of the economic progress which is deemed to have created it and also a form of global governance to replace nation state democracy which it is asserted is impotent to prevent it. The projection of humanity's suicide - so often contained in speeches to UN climate conferences and in those of gullible world leaders and heads of State[268] - is the effective means of maintaining eco-socialist propaganda.

[268] The latest being the *'suicide pact'* of the opening statement of the UN Secretary General to the 2022 CoP in Egypt.

It is the drum beat of the forecasts of catastrophe and condemnation of mankind that reverberates in Malthus[269], Jevons[270], Vogt[271], Carson[272], Marcuse[273], Erlich[274], Commoner[275], Bertrand Schneider & King[276], Stephen Schneider[277], Ashby[278], Brandt[279], Ward[280], Hansen[281], Strong[282], Houghton[283], Edenhofer, Leggewie and Schellnhuber[284] and now of Attenborough[285]

Set against the reality of an economy and of governance rooted in freedom and spontaneous response is the eco-socialist vision a monolithic structure of control and planning. Choice and variety through competition is to be overborne by collective monopoly.

Government close to the people, vulnerable to public opinion and sensitive to change is now to be vested in experts and remote authorities whose dictates cannot be questioned and for which their officials cannot be held accountable[286]. As is evident to all but the most obsessed the IPCC embodies this 'new order' in essential respects.

The collectivist view of humanity and its orderings is expressed not only in *les grand projets* but also in predictions of inevitable collapse of the existing order. All these are characterised in modern eco-socialism from Vogt to Schellnhuber and Attenborough.

Speed of fundamental change

It is the very speed of change that itself has challenged the capacity of humanity to assimilate it.

It is well known that evolution in Nature is measured in time by many generations and that the faculties of human beings to adapt to such change have not also evolved at the same rate or to the same extent. For at least 600 generations, since the formation of settlements of humans in social existence based on rearing of livestock and agriculture, human cognitive, emotional and moral faculties have been those which are adapted to individual survival and reproduction in very different circumstances and conditions. Those faculties are not readily adapted to a world utterly transformed by scientific discovery, technology and knowledge.

Steven Pinker, Professor of Psychology at Harvard describes how human cognitive and moral faculties are inherited from an 'archaic environment'[287]. In human cognition, superstitions remain potent forces prevailing over physics and biology: humans underestimate the extent of

---

[269] Thomas Malthus *An Essay on the Principle of Population* 1798
[270] William Jevons *'The Coal Question'*1865
[271] Vogt W.*Road to Survival* 1949 William Sloane Associates Inc
[272] Rachel Carson *Silent Spring* 1962
[273] Kellner D (editor) *Collected Papers of Herbert Marcuse* Vol 3 The New Left and the 1960s 2005 p 173 cited therein.
[274] Erlich P. *Population Bomb* 1968
[275] Commoner B *The Closing Circle* Random House Inc 1971
[276] Authors of *"The First Global Revolution"*1991 Pantheon Books.
[277] Schneider *Genesis Strategy* 1976 Plenum Press, New York. *"Global Warming"* 1989 Vintage Books.
[278] *'Pollution Nuisance of Nemesis'* A Report on the Control of Pollution 1972 p3 Chair Sir Eric Ashby FRS.
[279] Brandt Report re Commission on International Development 1980.
[280] Ward B. *Only One Earth: The Care and Maintenance of a Small Planet* 1982.
[281] Testimony to US Congress June 23 1988.
[282] Strong M. *Our Common Future* 1987 cited in Darwall R.
[283] John Houghton *"Global Warming: The Complete Briefing"*. Lion Press 1994.
[284] These three made their statement to the Essen Conference 2009 Darwall R. Green Tyranny 2019 Encounter Books pp 15 – 22.
[285]Attenborough D. *"(We) are on the verge of destabilising the entire planet"*. Lee, Dulcie; Lee, Joseph (13 June 2021) Wikipedia . *"Humans are a plague on the Earth"* Radio Times. 22 January 2013 Wikipedia. See also BBC podcast by Attenborough 7 October 2020 cited in Crockford S OpCit p125/126
[286] The German eco socialist Schellnhuber and adviser to its government demands that judges be empowered to overturn the will of a democratic majority. Darwall R. Green Tyranny Op Cit p16,17. He stated 2017 that if no was action taken by 2020, the world *"may be fatally wounded."*[286].
[287] Professor Pinker S *Enlightenment Now* 2018 Penguin Random House pp25,26

coincidence: infer causation from correlation: think in black and white: generalise from anecdotal or private experience: rationalise by stereotype: project conceptions of typical traits of a class or group on to any individual deemed to belong to it: marshal evidence confirming their convictions while dismissing evidence that contradicts them.

In the moral faculty also humans act in ways that derive from an earlier and discretely different age. Thus people demonise those who do not agree with them: attribute differences of opinion to stupidity and dishonesty: seek scapegoats for every mishap or misfortune: employ moral grounds for condemning those with opposing views: mobilise moral indignation against them including for having questioned authority or flouted custom or undermined tribal solidarity.

The bitter hostility to those who reject or cast doubt on global warming dogma conforms to each of these characteristics that Professor Pinker lucidly identifies.

Effect and Intent

It is salutary to apply a test of legal accountability as to the intent of the eco-socialist assault on the open market economy and nation state democracy. It is an axiom. The deducing of the intent from the effect. If a given outcome is certain to any rational being to be the consequence of a voluntary act then it is presumed conclusively to be its intent, in the absence of evidence of a new and effective cause. The outcome of eliminating fossil fuel combustion will be the collapse of the open market economies of those States that enforce it.

What other States will benefit from such a disaster? Surely they are China and Russia both of whom give no credence to the global warming deception.

How have the liberal democracies of the western world allowed Strong's imperative of the collapse of industrialised civilisation[288] to become even a possibility?

How is it that such a dismal and fallacious conception of a fearful and imperilled humanity has been allowed to paralyse reason?

How is it that fundamental laws are denied by a form of obsessed belief and policy rooted in a monstrous deception is dictated by an unaccountable remote bureaucracy?

Such questions were addressed by Hayek in 1944 when western Europe was convulsed in a war with a German totalitarian regime and about to become a combatant in a 45 year Cold War with a Russian tyranny. Hayek's *Road to Serfdom*[289] is a short book. It sets out the essential case for freedom and its incarnations in forms of economic life and political governance. It is a paramount testament to the blessing for humanity that is freedom. The following Sections of this Part 6 examine the essential distinctions of the open market economy in a liberal democracy and the collective organisation of production under a socialist form of governing bureaucracy .

[288] Maurice Strong, Interview 1992,concerning a book he hoped to write. Gibson, Donald. Environmentalism: ideology and power. p 95 (Wikipedia.)
[289] Hayek F. *The Road to Serfdom* 1944 Routledge.

II. Free society and the individual

Interaction and unpredictability

Hayek explains that forming the root of a free society is the interaction of humans in a spontaneous, unpredicted way with unrestricted exchange. It is an interaction of individuals, possessing different knowledge and different views and it constitutes the 'life of thought'[290].

The growth of reason, Hayek contends, is a social process based on the existence of such differences, their resolution and further interaction. Its results cannot be predicted. We cannot know which views will assist this growth and which will not – in short, this growth cannot be governed by any views which we now possess without at the same time limiting it[291]. By attempting to control the interpersonal process, to which the growth of reason is due, we are merely setting bounds to its development and must sooner or later produce stagnation of thought and decline of reason.

The principle of the individual

Hayek's exposition[292] of the origin of a free market economy is one of historical evolution, by countless individual spontaneous responses, from the seed bed of the Renaissance that asserted the dignity and value of the human individual which it had derived from classical philosophy[293]. That individualism flowered into what we regard as Western European civilisation. Respect for the individual man *qua* man: the recognition of his own views and tastes as supreme in his own sphere - however that may be circumscribed: the belief that that men should develop their own individual gifts and bents.

Perhaps the most beneficial for humanity of the unchaining of individual energies was the growth of science which followed '*the march of individual liberty from Italy to England and beyond*'. Only since industrial freedom opened the path to the free use of knowledge, only since everything could be tried – if somebody could be found to back it at his own risk-- has science made the great strides which in the last 150 years have changed the face of the world.

Fluidity and freedom

There is nothing in such liberalism to make it a stationary creed: there are no rules fixed once and for all. The fundamental principle is that, in the ordering of our affairs, we should make as much as possible of the spontaneous forces of society and resort as little as possible to coercion. That principle is capable of an infinite variety of applications. These include deliberately creating a system in which competition works as beneficially as possible and not slavishly applying rules of thumb - including *laissez faire.*

Hayek concludes that liberal democracy can only flower within economies based on a competitive system of freedom to dispose of private property in all its forms, by whatever means.

---

[290] Hayek. F. *Road to Serfdom* p 169.
[291] "The tragedy of collectivist thought is that while it starts out to make reason supreme, it ends by destroying reason because it misconceives the process on which the growth of reason depends. Individualism ...is an attitude of humility before this social process – of tolerance to other opinions and is the exact opposite of that intellectual hubris which is at the root of the demand for comprehensive direction of the social process." pp169/170 Hayek. *Road to Serfdom.*
[292] F.A Hayek Op cit Ch 1 pp13 – 16.
[293] "*the basic individualism inherited by us from Erasmus and Montaigne, from Cicero and Tacitus, Pericles and Thucydides* Op cit Ch 1 p 14.

A collectivist controlled economy requires the constraint of disrupting dissent[294].. Equally democracy is an obstacle to the coercion and suppression of freedom which the direction of economic activity requires.

Democratic restraint on power

It is not the fact that democracy has the will of the people as its authority that guarantees resistance to arbitrary power. It is not the source of power but the limitation of power that is inherent in democratic control which may enable it to prevent power becoming arbitrary[295]. It is the price of democracy that the possibilities of conscious control are restricted to fields where true agreement exist for, as Hayek asserts, *"in a society which for its functioning depends on central planning this control cannot be made dependent on a majority being able to agree"*

That factor accounts for demands by eco-socialists that the will of a small minority be imposed upon the people since it is this minority that forms the largest group able to agree amongst themselves. It accounts for demands that experts[296] be empowered to dictate the drastic changes in economic and social order deemed necessary to meet the 'challenge of climate change'.

The force of such a principle extends to the de-industrialisation of an entire country consequent on the elimination of the only viable energy base on which its economy depends. The evidence of the past 30 years does not give any solid assurance that ultimately the fundamentalism of eco-socialism will not overwhelm democratic restraint on bringing that about.

Let Richard Feynman[297] have the last word on socialist fundamentalism which he describes as:-

> 'The antithesis of the scientific method since in [such] socialism the answers are given to all questions without discussion and without doubt. The scientific viewpoint is the opposite of this. We must argue everything out, observe things, check them and so change them. Democratic government is much closer to this idea. There is discussion and the chance of modification. One does not launch a ship in a definite direction. It is true that if you have a tyranny of ideas - so that you know exactly what has to be true - you act very decisively. But soon the ship is heading in the wrong direction and no one can modify the direction any more. So the uncertainties of life in a democracy are much more consistent with science.'

[294] F.A Hayek Op cit Ch 5 p 73.

[295] F.A Hayek Op cit Ch 5 p 74.

[296] See statements of Club of Rome's 3 reports and of Strong, Edenhofer, Schellnhuber, Leggerwie and others referred to in Part 5.

[297] *Essay in The Pleasures of Finding Out*". Richard Feynman 1918 – 1988 American theoretical physicist who was regarded as the most brilliant, influential, and iconoclastic figure in his field in the post-World War II era. He re-constructed quantum electrodynamics—the theory of the interaction between light and matter—and thus altered the way science understands the nature of waves and particles.

III. Functioning of the open market economy

The open market economy has grown, by countless individual spontaneous responses, as a form of mutual beneficial collaboration within a framework of rules or of an order evolved through adaptation. Such an order, in its essentials, must secure private property, allow a system of credit and enable coordination of skill and knowledge by competition and, by the concepts of price and profit, ensure information, economic allocation of resources and efficiency.

Private property

The notion of private property is misleading insofar as it connotes property which may conversely be owned by the public. The property held nominally in the United Kingdom by the organs of state is simply property vested in persons having public official duties. It is held by a single ultimate entity and thus by distinction private property is sometimes referred to as 'several' property – including by Hayek.

Freedom to pursue distinct aims impelled by differing skills and knowledge, depends on a general acknowledgment of the existence of a private domain of which the individual is free to dispose. It also extends to the manner in which such right may be exercised to transfer property in things.

> *"The prerequisite for the existence of such property, freedom and order from the time of the Greeks to the present is the same: law in the sense of abstract rules enabling any individual to ascertain at any time who is entitled to dispose over any particular thing.[298]"*

It is the abiding achievement of the English legal system that it still comprehends a system of common law and equity[299] developed by precedent but adapted to conditions and circumstances of changing times and values. It represents, in the realm of the rule of law, evolution of an order by spontaneous[300] response to the realities of human existence and reached its most unrestricted flowering in the age of the Industrial Revolution of the 19th century.

Credit

Adam Smith, the founding spirit of the free and open market economy, declared that "*Money is belief*". It is a trust or faith. Essentially it is credit, its root verb being the Latin 'Credo' – the 'I believe'.

The two essential factors required for the operation of an free market economy are assumption of Risk and the extension of Credit. Ownership of the means of production, distribution and exchange is not essential. Nor is labour - as the advance of technology has demonstrated.

---

[298] Hayek F *The Fatal Conceit* 1988 Chicago University Press pp 31,32.

[299] Legal remedies other than damages, such as writs, injunctions, and specific performance allowed the chancellor to decide the most equitable course to take in each individual case.". It was systemised by Lord Eldon in the early 19th century.

[300] Occurring without external constraint or impetus. Latin *(sua) sponte* = willing, of one's own accord.

Reward for successful risk taking enlarges credit and failure diminishes it[301]. Credit is trust in economic form. Each such term has the same meaning and root. All start-ups require trust since security in the form of accumulated credit cannot be provided.

Risk is the willingness to embark on what cannot be fully forecast in order to effect the supply of what is perceived may be useful or necessary to others.

All this is reflected in the rapid and wide extension of banks and stock exchanges in Great Britain as the Industrial Revolution developed so facilitating credit and investment..

Central to the operation of an open market economy are competition, price and profit.

Competition

Competition among consumers through prices draws finance, labour and other resources to flow to unsatisfied demand which is where rates of return are highest. It is also competition of producers which then equalises those returns. Thus competition ensures sharing of resources through pricing.

Informing the Maurice Strong eco-socialist view of a moral global governance is the notion of 'co-operation'. It is attractive as an ideal, suggesting absence of conflict. However in common with other terms of 'social' theory it has no useful meaning as a principle of economic order. It is similar to "solidarity". It presupposes a large measure of agreement as to its ends. It also assumes a consensus as to the means to be employed to achieve them.

Whilst this may be achievable in a small group of people having like attitudes, habits, knowledge and beliefs it affords no basis for adapting to the conditions and circumstance of what cannot be predicted with certainty.

The dominant and immutable fact is that the future is unknowable and that change is of the very nature of the manifest world. Accordingly, adaptation to the unknown is an indispensable process and is that which permits evolution to occur in an economic order. The necessary coordination of effort and knowledge for this adaptation is afforded by competition.

> *"It is a procedure of discovery, a procedure involved in all evolution, that led man unwittingly to respond to novel situations: and through further competition, not through agreement, we gradually extend our efficiency. Neither all ends pursued nor all means used are or need to be known to anybody in order for them to be taken account of within a spontaneous order".*[302]

The beneficial operation of competition requires that its participants observe rules to unite an evolving extended order. Hayek explains that "*such an order forms by itself*". It arises by the adoption of rules that are adaptive. It is an evolution that is not linear but results from continued trial and error, constant experimentation.

[301] It was the collapse of Credit-Anstalt in May 1931 that set off a hain reaction of loss of confidence in credit triggering run on German banks, withdrawals in London, devaluation of sterling and US bank failures. The financial crisis of 2007/8 was triggered by collapse of credit availability and elimination of liquidity in the banking sector.

[302] Hayek F The Fatal Conceit 1988 Chicago University Press pp 19,20.

Profit and Price

Eco-socialists juxtapose pursuit of profit with notions of 'private greed' and' public good'.

Altruism[303], like many virtuous concepts, is one which has no utility in reality since there is no means by which it can in practice be determined sufficient to ground common action. Producers, distributors, traders and financiers are not governed by altruism. But it is the very concern for profit that makes possible the most effective use of resources. It is by the impelling necessity of profit that productive capacity is multiplied by different individuals obtaining access to different knowledge whose total exceeds what any single one of them could muster.

Prices and profit are all that most producers need more effectively to serve the needs of people they do not know[304]. They are a tool for searching so that with the market process it is possible for people to be given the information and material resources they need to obtain the object of it all – that they get what they want.

At Thomas Sowell explains with his customary simplicity and perception[305] a free open market system is one which prices act as a fast and effective conveyor of information through a vast society in which fragmented knowledge is coordinated. It is not a profit system. It is a profit-and-loss system. Losses are of equal importance for the efficiency of the economy as losses stop the practice of putting resources into what customers do not want - either by voluntary action or bankruptcy. From the standpoint of the economy as a whole, society is using its resources, which are necessarily 'scarce,' more efficiently because decisions are guided by price.

The staggering amount of economic transactions, on ever changing terms as supply and demand vary continuously, is beyond the knowledge and capacity of any individual or any group of planners to direct. Moreover each of the billions of people involved in market transactions around the world need concern themselves only with what are their own individual transactions.

As noted above when more of some item is supplied than demanded, competition among sellers to get rid of the excess will force the price down with resources for that item being set free for use in producing something else in greater demand. Conversely when demand for an item exceeds supply rising prices due to competition among consumers encourages more production so drawing resources away from other parts of the economy to accomplish that.

Efficiency of allocation of resources governs the prosperity of entire countries. This is demonstrated by the experience of China, India and Venezuela. Since China and India adopted open markets systems GDP per capita has risen since 1995 by a multiple of 20 in China ($609 to $12,556) and by 6 times in India ($374 to $2,280). In the past 12 years of far-left socialist rule in Venezuela GDP per capita has collapsed from $12,180 to $1,823 – despite its vast oil resources.

---

[303] Hayek F The Fatal Conceit 1988 Chicago University Press p 104.

[304] Hayek was awarded the 1974 Nobel Memorial Prize in Economic Sciences for his work showing how prices communicate information.

[305] Thomas Sowell *Basic Economics A Common Sense Guide to the Economy* Perseus Books Group NY 2011 Ch 2 The Role of Prices pp 11- 39 for the complete text.

IV. Eco- Socialism. Propaganda. Freedom

Part 5 described how it was that eco-socialism adopted and exploited the dogma of global warming to justify demands for fundamental changes to economic and political structure that are inimical to the free open market economy.

Such is the rising tide of anti-capitalism that is beginning to extend into the prevailing culture and consensus of opinion of today - and so significant is this development that it warrants an enquiry as to how it has come about. Propaganda[306] as a tool of deceit has been examined in Part 2 in the context of the IPCC. It is also by well conceived propaganda repeatedly disseminated that eco-socialism has achieved the widespread submission of free democratic peoples to its dogma of catastrophe and need for revolutionary change.

Eco-Socialist Networks and Propaganda

It is imperative to anyone who values freedom of speech to realise the vast and dense network of institutes, advisory bodies, government departments, supranational entities[307], foundations, trusts and corporations having direct influence on policy and resource allocation for 'meeting the challenge' of climate change and promoting the eco-socialist dogma of global warming.

The climate change cult network in the UK alone is of extraordinary complexity and extent[308]. An example of its sinister grip on advice and policy as to climate change and resource allocation is the case of one such environmentalist Michael Grubb.

His website lists him as having directed the Chatham House[309] Energy and Environment programme for 10 years prior to his appointment as Professor of Energy and Climate Change at University College London and Research Director for the 'Institute of Sustainable Resources & Energy'. On the UCL Masters Course, 'Economics and Policy of Energy and Environment', he disseminates the dogma to students from around the world in a course described as a *'core module on the economics and political economy of energy and climate mitigation policy'*.

For the past two years he has taught Planetary Economics also at Beihang University in Beijing.

He was founder and Editor-in-Chief of the 'Climate Policy' journal which propagates global warming theory. From 2008 to 2011 he served on the UK Climate Change Committee, formed under the UK Climate Change Act 2008 to enforce the UK carbon dioxide budget.

From 2011-2016 he was engaged at the UK regulator Office of Gas and Electricity Markets (Ofgem) as Senior Advisor on 'Sustainable Energy Policy' and 'Improving Regulation'. In 2016 he became chairman of the UK government 'Panel of Technical Experts on Electricity Market Reform'. In 2018 he was appointed as Convening Lead Author for Chapter 1 of the IPCC Sixth Assessment (Mitigation Report). In the same year he was appointed as 'Leader for the Sustainability' hub of the UK Research Council's programme on 'Rebuilding Macroeconomics'. He

---

[306] *Propaganda* 'that which is to be propagated 'ablative singular feminine of the gerundive form of *propagare* 'to propagate', 'spread' 'disseminate'.[4]

[307] World Bank. European Commission. Intergovernmental Panel on Climate Change (IPCC). World Health Organization (WHO) United Nations Development Programme (UNDP) United Nations Framework Convention on Climate Change (UNFCCC) UN Office for Disaster Risk Reduction (UNFDRR) United Nations Environment Programme (UNEP).

[308] See a brilliant analysis of Rupert Darwall in *Green Tyranny* 2017/19 OpCit at pp175 – 178.

[309] Royal Institute of International Affairs, Chatham House London is a policy institute with a stated mission to provide commentary on world events and offer solutions to global challenges.

is on the Scientific Advisory Boards of the German Institute for Economic Research with its remit on climate change[310].

Before joining UCL he was also a Senior Research Associate in Economics at Cambridge University and, prior to joining Ofgem, was Chief Economist at the Carbon Trust and Chair of the international research network and interface organisation 'Climate Strategies' whose mission statement asserts that *"Climate change is our greatest planetary threat, affecting all species and ecosystems"*.

The Climate Strategies website includes a statement that it is a *"world class independent policy and research input to European and International climate policy"*. Its list of collaborators and supporters reveals the extent of the arteries and veins through which eco-socialist dogma now flows. It includes the UK government Foreign and Commonwealth Office, Dept of Energy and Climate Change, Dept for International Development and Committee on Climate Change, the German Federal ministries of Economic Affairs Energy and the Environment, Nature Conservation and Nuclear Safety as well as a number of foundations and institutes.

As has been lucidly described[311] the 30 years of allocation of truly vast resources to the dogma of global warming – larger than any in history – is sustained by a dense network of people defined by the same outlook, committed to the same goal and – crucially – agreeing on the means of achieving it. It is a monopoly of expertise and advice to governments, creating a culture in which scepticism is a capital offence since it poses fundamental questions which would cast doubt over the whole enterprise. The echo chamber created by wall to wall climate propaganda is mirrored within government bureaucracies. The 'experts' they commission and the advisers seconded to them are all drawn from the same milieu.

It is surely wicked that those holding themselves out as scientists should persist in profiting from the dissemination of eco-socialist 'climate change' dogma knowing as they must or ought to do that CO_2 cannot as a matter of physics cause rise in global warming and never has done[312].

Hayek's warnings as to Propaganda

Hayek presents an analysis of the means by which propaganda is most effectively deployed. He provides a compelling description of the essentials of propaganda to support a single end or system of pre-determined outcomes, deemed to be for the benefit of society and morally virtuous.

His observations, when applied to the dissemination of the doctrine of man made global warming, attain a fearful reality for our times. They provide a context for examining the eco-socialist nature of global warming propaganda.

Public acquiescence and acceptance

Hayek explains that the most effective way of making everybody serve the ends towards which a programme for change is directed by a higher authority is to make everybody believe in those

[310] Its **Department Climate Policy** is described as using *"empirical and theoretical methods to explore Germany and Europe's transition to sustainable energy"*

[311] Darwall R. *Green Tyranny* Encounter Books 2019 pp 175/7 from which this text is derived. It is a lucid and comprehensive survey.

[312] It has saturated all available radiation at levels of concentration 66% of CO_2 current levels. More CO_2 has negligible effect. See fully in Part I

ends. It is essential that the people should come to regard them as their own ends. The beliefs may be chosen for the people and imposed upon them but they must nevertheless become their beliefs: a generally accepted creed which makes the individuals act spontaneously yet in the way that the authority requires[313].

Hayek observes that, in addition, this will be best achieved by persuading the people that the validity of the values they are to serve are really the same as those which they have always held, but which were not properly understood or recognised before[314].

The fact that Vogt's *Road to Survival* and the Club of Rome's *Limits to Growth* sold millions of copies is testament to the appeal they made to basic insecurities of humans instilled by fear of change and fear of scarcity. A depiction of nature as a fixed and limited 'resource' was readily embraced by a society witnessing sudden a profound revolution.

But the nature of 'growth' was fundamentally misconceived. It is not growth that uses resources but production and consumption. Growth is simply the outcome of application to resources of human imagination, intelligence and knowledge. The advances in science and technology have allowed humans to abandon less efficient or less available resources for more efficient and more accessible ones. Moreover growth is now, to a vast extent, attributable to the evolution of the digital age in which economic yield is not dominated by finite resources in a fixed system.

Growth is inherent in nature and underwrites the evolution and development of organisms. Growth in this sense is a process occurring in a self-maintaining structure[315] .

There is no evidence that the Earth's resources- as distinct from reserves - will be insufficient for our needs for the foreseeable future – measured in thousands of years. It is a logical extrapolation that has no validity in reality. We cannot form useful conclusions as to the human population, brain power, innovative discovery, intelligent response, and adaptation to change over the next 100 years much less over 1000 years.

Suppression of opinion, evidence and doubt

It is cardinal condition of effective propaganda both that the people are converted willingly to the dogma and also that the gospel itself must not be exposed to rational enquiry. It must be declared as an inviolate belief. Its creed must become sacrosanct and utterly exempt from criticism.

"Public criticism or even expressions of doubt must be suppressed. The whole apparatus for spreading knowledge, the schools and the Press, wireless and cinema, will be used exclusively to spread those views which ... will strengthen the belief.. and all information that might cause doubt or hesitation will be withheld"[316].

So spoke Hayek in 1944. What more compelling confirmation could there be of his insight into propaganda than the cynical adoption of its precepts by the founders of modern eco-socialism? Is not the indoctrination of children one of the most wicked of the injuries inflicted by its

---

[313] Hayek F *Road to Serfdom* 1944 Routledge p 157
[314] F Hayek *Road to Serfdom* 1944 Op cit p 161
[315] F Hayek *The Fatal Conceit* 1988 p144 - 146
[316] F Hayek *Road to Serfdom* 1944 Op cit p 164

adherents? How is it that the fundamentalism of global warming has gained such supremacy over free expression, doubt and scepticsm that have hitherto been the foundations of the scientific method in our tolerant and open society?

"The word truth itself ceases to have its old meaning. It no longer describes something to be found with the individual conscience as the sole arbiter of whether...the evidence warrants a belief: it becomes something to be laid down by authority. Which has to be believed in the interest of the organised effort"[317]

Is this not an indictment of eco-socialist global warming propaganda?

Who can honestly affirm that they have not acquiesced in the loss of something priceless?

A regime of suffocation of free expression of opinion has stifled enquiry and doubt. Contradictory opinion or evidence fails not only to be publicised in the organs of news and public opinion but is also suppressed even when uttered by leading professors of physics and of other disciplines bearing directly on thermal radiation and atmospheric gasses.

It is now the case in England that all organs of broadcasting and sources[318] of current information,[319] with one exception,[320] are unanimous in their submission to the dogma of global warming and the gospel of Net Zero

Once this happens then, as Hayek explains,

> *"It is no longer a question of persuading the people ...The skillful propagandist then has the power to mould their minds....and even the most intelligent and independent people cannot entirely escape that influence if they are long isolated from all other sources of information.*[321]*"*

Dogmatic suppression of opinion in the name of belief in incontrovertible truth is the watermark of the religious fanaticsm. Eco-socialist dogma engenders the same obliterating intolerance[322]. and even hatred of those who resist its claims.[323] Only the fanatic would deny this.

Ignorance of the truth is voluntary

What is it that restrains people from seeking the truth about the Earth's temperature?

It is open to anyone to click on Dr Roy Spencer[324] on Google search engine in order to reveal the precise variations of the temperature of the Earth for each of the 44 years since satellite and

---

[317] F Hayek *Road to Serfdom* 1944 Op cit p 167

[318] Press statements of Guardian columnists *'climate change denial now looks as stupid and as unacceptable as Holocaust denial'* G Monbiot September 2006 *'climate change may be an issue as severe as war. It may be necessary to put democracy on hold for a while'.* James Lovelock May 2010,

[319] Wikipedia, the world's most influential information source, excludes articles explaining the flaws in global warming theory. It describes those who question its validity as guilty of *'denial, dismissal, or unwarranted doubt that contradicts the scientific consensus on climate change'* and describes the contra arguments as simply *'manufactured uncertainty'*

[320] GB News permits commentary as to the folly of Net Zero but has not yet exposed the fundamental fallacy in the light of the principle of CO_2 saturation with electromagnetic radiation

[321] Hayek F *Road to Serfdom* 1944 Routledge p 158

[322] [5]Charles then Prince of Wales describes those casting doubt on the hypothesis as "headless chickens". Ed Davey Sec of State for Climate Change described them as "wilfully ignorant".

[323] Nigel Lawson former Chancellor of the Exchequer wrote "An Appeal to Reason" in 1998 arguing that 'consensus' claimed by the IPCC is not a valid basis for science which is far from settled.. He has spoken publicly of the enmity towards him following publication. *"I have never in my life experienced such extremes of personal hostility, vituperation and vilification."* Speech University of Bath 26 April 2014.

[324] University of Alabama Huntsville USA.

balloon radio sonde measurement became continuously available – Figure 28 at page 68.

All of the fluctuations in temperature shown in the satellite record are accounted for by variations of solar activity during its 11 year cycles as well as the effect of massive naturally occurring oceanic shifts of temperature and our emergence from the Little Ice Age.

Figure 29 below literally illuminates the effect of the presence and absence of solar activity.

Figure 29

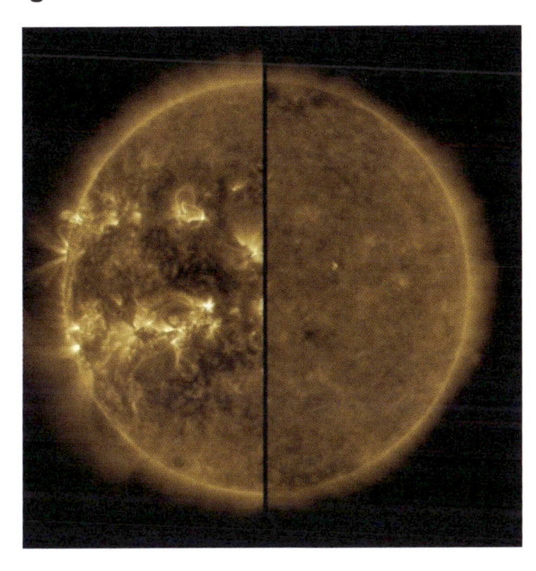

This split image shows the difference between an active Sun during solar maximum on the left and a quiet Sun during solar minimum on the right.

With the commencement of each cycle the Sun's magnetic activity intensifies until the solar maximum for that 11 year cycle..

How is it that with the unimpeachable evidence of satellite observations over some 43 years it is possible that the charade of global warming is still believed as reality?

Asserting, correctly, today that the laws of physics falsify the dogma of global warming is similar to the rejection by the Lollards 650 years ago of the dogma of transubstantiation– belief in the metamorphosis of bread and wine into the actual flesh and blood of Christ. Statements of the manifest empirical fact that this could not occur were in 1382 declared to be heretical with those guilty of making them excommunicated and some being burned to death.

Global warming dogma gains acceptance with not a murmur of public outcry as to the colossal deception involved in its dissemination. Such is the power of propaganda in suppression of contradictory evidence and dissent.

As Hayek observes propaganda undermines *"the sense and the respect for truth."*[325] Its proponents *"will readily embrace theories which seem to provide a rational justification for shared prejudices. Thus a pseudo-scientific theory becomes part of the official creed which to a greater or lesser degree governs everybody's action".*

Such a passage could have been written specially as a comment on the fraudulent misrepresentations in the diagrams in each of the IPCC 2001 and the IPCC 2021 Summary for Policymakers as to historical temperature levels alleged to have been detected by proxy analysis (see Part 2 page 13).

[325] Hayek F *Road to Serfdom* 1944 Routledge p 161.

Distortions of meaning

The use of words with negative connotations displacing previous meanings is seen by Hayek as a very efficient technique in procuring acceptance of the values of the prevailing propagated dogma.

Confucius put the matter simply.

When words lose their meaning

People will lose their liberty

Just a few examples of the eco-socialist distortions of meaning illustrate Hayek's perception.

'**Growth**' is now a pejorative term signifying reckless consumption of finite resources.

'Liberalism' now connotes socialist conceptions of the basis of economic and political order.

'**Denier'** becomes an accusation of heresy, bigotry and even criminality.

'**Environment'** no longer means the immediate natural world but extends to atmospheric gasses that it is contended 'pollute' and 'harm' the planet.

'Natural' excludes the evolutionary process of adapting and developing in a self-sustaining way which characterises mankind just as much as the finches of Galapagos.

'**Capital'** does not connote available credit for investment but of ownership and control of a store of wealth limited to a class of few who control the process of production to the exclusion of the many who do not have such ownership and control. Those excluded are classified as Labour.

"**Capitalism**" is as false a term as what it is intended to describe. The open market economy depends on maintained credit and the assumption of risk. 'Capitalism' was not a term known to Marx in 1867 when he completed Das Kapital. He never used it. It emerged in political debate as the opposite of socialism only with publication of *Der Modern Kapitalismus* in 1902[326]. The creation of the notion of a special class interest group of 'capitalist' owners of accumulated earnings from production ('capital') ensured that is would be opposed by the working proletariat[327] who were in reality its main beneficiaries. It was, in truth, the free market economy and risk investment that enabled the working poor at first to survive and then to rise to a level of prosperity unimaginable two generations before.

The tragic dualism of 'Capitalism' and 'Socialism' has riven society and the peace of the world. It is the concept of the adversary. It denies the natural justice of the fluidity, spontaneity and innovation of a free market economy sustained by credit, confidence and opportunity.

---

[326] Sombart W. *Modern Capitalism* 1902 Braudel 1982a 277 cited at p 111 F Hayek The Fatal Conceit Op cit.

[327] F Hayek The Fatal Conceit Op cit p 111.

V. Eco -Socialism v. Free market economies

The emergence of eco-socialism and its subversion of environmentalism with global warming dogma has been examined in Part 5 above. Its demand for overthrow of the open market economy requires the abandonment of the reality of fluid, ever changing, responsive and limitless interactions of individuals as they lead their lives in ceaseless flux, in changing conditions and degrees of prosperity and opportunity.

Dualism and disintegration

The eco-socialist assault on this the very basis of an unprecedented accession of human welfare rests on both the creation of an enemy and also on the imposition of a conceptual master matrix that explains everything, bears no contradiction and denies mankind his essential humanity depending as it does upon freedom and choice.

The consequences of this are manifest and undeniable. The shame of death and destruction wrought by socialist dogma in China, Cambodia, Soviet Russia and the tyrannies of Eastern Europe in just 70 years of the 20th century should have borne conclusive testament to its wickedness.

Such is not the case. Socialism now flourishes as the dominant ideology of 'climate change' as has been demonstrated in Part 5.

Socialist dogma requires that there must be an enemy and victim[328], oppressor and the oppressed, despoiler and the despoiled, the few and the many, the rich and the poor. Classification of human beings is a vice of Marxism. It is a duality that promotes dis-integration and rejects integration. It fosters separation not wholeness. Nor does there seem to be any limit to the extent of conceived victimhood, guilt and fear arising from its dissemination.

There is irrefutable evidence that eco-socialism is has entered the arteries of the political culture of Western democracies and that its ideology has infected the main organs of opinion, broadcasting, education and the Press.

These are not idle speculations. They go to the root of the eco-socialist crusade against Man[329] and the economic and political structures that brought about the miraculous transformation of human existence. Yet no class or category of humans in society is rigid, self-contained, discrete and readily identifiable by empirical observation. Human beings are not so confined in a free society - as individuals they move, adapt, innovate, evolve, endure, rise, fall and change.

Global governance

Eco-socialism envisages global authority overreaching national self-governing democracies. As noted above Hayek declares that democracy clashes with such totalitarianism. Democracy is the *"obstacle to the suppression of freedom"* which its imposition would require[330] not by reason of

[328] *Hayek Road to Serfdom* 1944 George Routledge & Sons p143

[329] The extreme circular drafted by a directors of a dedicated German environmentalist group issued in 2015 by the 87 year old head of the Christian catholic church *Laudato si'* condemns humanity for causing dangerous change to the climate and acting contrary to God's wish in his 'destruction' of the natural environment. See Parts 5 and 7 for complete rebuttals of these absurd statements.

[330] Professor Hayek Road to Serfdom Ch 5 p74.

its authority being derived from the people's will but rather by the nature of its processes.

Liberal democratic systems are condemned by apostles of the eco-socialist new order as being far too slow to act, too sensitive to the flux and sensitivities of public opinion, changing events and circumstances. However if individual liberty is indeed to take precedence then it would not be possible for the whole of society and all its resources to be organised for a unitary end[331].

Eco-socialists assert that global governance structures are *"the indispensable driving force"* of the necessary transformation and that climate change objectives must be embedded in national constitutions.[332] *"Climate change means change of culture requiring a new economic order, new instruments of global governance[333]"*

No greater unitary end can be envisaged than 'de-carbonisation' for it would bring about de-industrialisation – a revolution so catastrophic that it could not be inflicted by rational and informed human beings on each other.

Fragile as it may be and imperfect as it is, democracy is our assurance of freedom for so long as it fosters resilience, contrary opinions, open discourse and the clash of ideas.

We owe it to those who are yet unborn to expose, confront and arrest the further descent into a form of society, of government and of an economy envisaged by adherents of the cult of eco-socialism before irreversible harm is inflicted.

SUMMARY

The root of a free society is the exchange of knowledge and ideas by interaction of individuals in spontaneous, unpredicted responses. Only with freedom of knowledge and industrial opportunity - only since everything could be tried and somebody found to back it at his own risk-- has science changed the face of the world. Democracy is the obstacle to the loss of freedom which eco-socialism necessarily requires.

---

[331] Professor Hayek *Road to Serfdom* Ch 5.

[332] See Germany's Advisory Council on Global Change "*World in Transition. A Social Contract for Sustainability: Summary for Policy Makers* (2011) cited by Rupert Darwall *The Green Tyranny* Encounter Books 2011 p22. Also Potsdam Institute for Climate Impact Research a German govt-funded institute.

[333] Article Professor C Leggewie Die Welt May 2011 cited by Rupert Darwall Op cit p 21 . See also C. Leggewie and H Welzer *The End of the World as We Know It: Climate, the Future and the Prospect for Democracy* 2009 cited by Darwall p21.

Part 7

THE SUNLIT UPLANDS

"The life of the world may move forward into broad, sunlit uplands".
Winston Churchill. House of Commons 18 June 1940.

The prediction made by Winston Churchill of the outcome of victory over tyranny came true.

The entire world lives in a golden age save for regions afflicted by war, civil unrest or tyranny. Not one of the predictions of doom and catastrophe that were made as to the impacts of mankind in Nature were true. All have been falsified by events.

'Silent Spring' claims of pesticide carcinogens inflicting mass deaths were false. Forests were not being destroyed by acid rain. There could not be any nuclear winter. There was no 'hole' in the atmosphere. There was no ice age. There has not been nor can there ever be an increase in the temperature of the Earth not accounted for by solar activity, Earth's orbits and ocean shifts.

Over the 82 years since Churchill uttered his clarion call to the defence of freedom humanity has risen to a height of welfare, prosperity and security that defies comprehension. There is no historical context for such rapid and all-embracing advance in beneficial existence.

Humanity's ascent to the broad sunlit uplands in just 4 out of the 8,000 generations since *homo sapiens* emerged is none the less capable of being measured with objectivity and precision.

I. LIFE EXPECTANCY

The following graph depicts the life expectancy of world populations from the earliest emergence of the industrial revolution in England in 1770 to 2019.

Figure 30

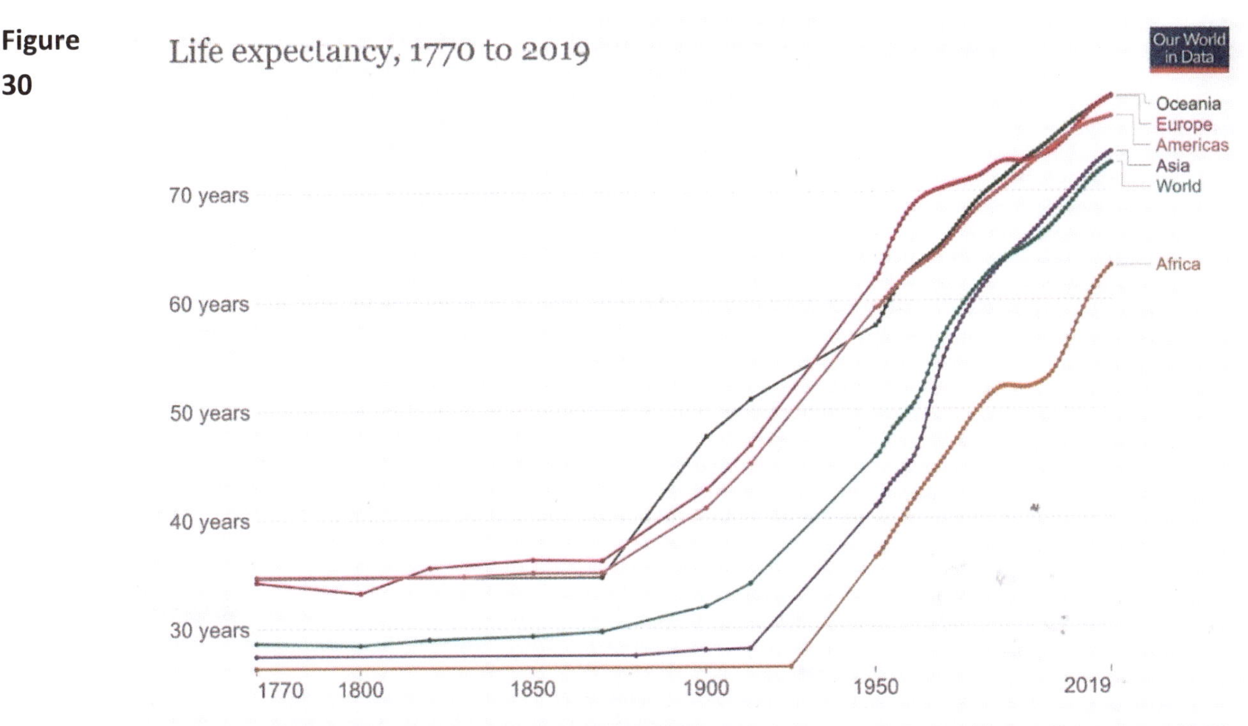

In the last 200 years populations of most of the richest countries in the world have more than doubled. By 2019 the life expectancy in Spain, Switzerland, Italy, and Australia was over 83 years. In Japan it was 85 years. By 2019 in African countries having poor medical care, sanitation and nutrition the worst health life expectancy was nevertheless between 50 and 60 years - the lowest being that of the population of the Central African Republic at 53 years.

300 years before 1850

Since at least 1550 until 1850 people living in the Americas and Europe could not expect to live more than 35 years. For the World the average expectancy was just 29 years.

Figure 31

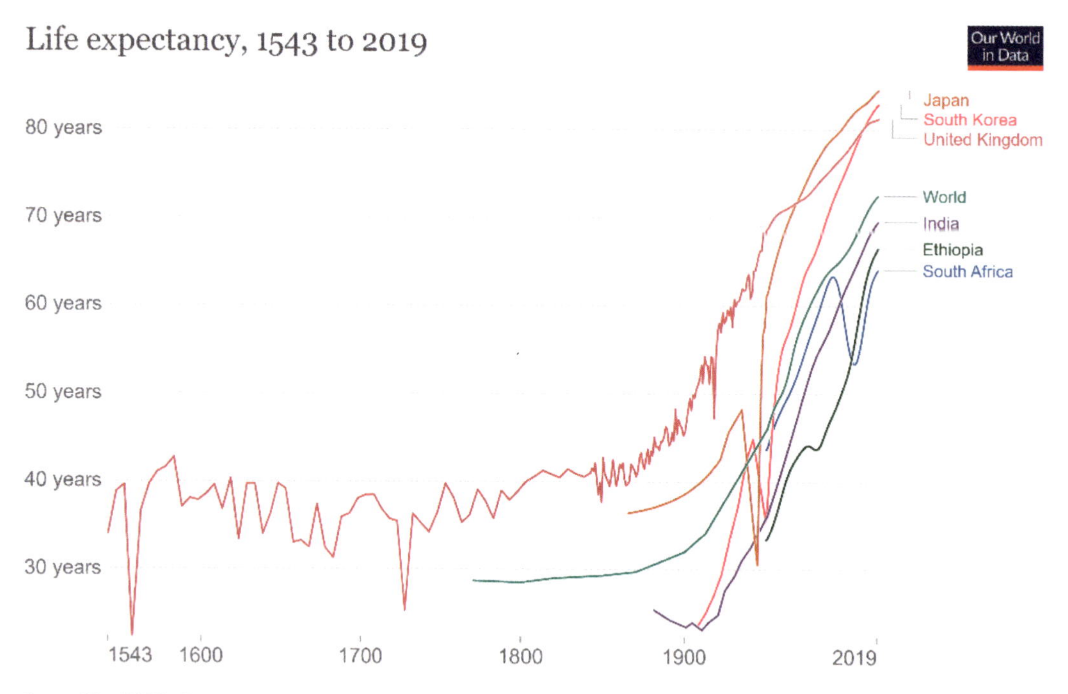

Moreover that had been the range of expectancy for almost all of human existence. Hunter gatherers had a life expectancy of some 32.5 years. With the advent of settlements and farming it is thought that that life expectancy fell due to the change of diet to carbohydrates and to disease but rising again in the Bronze Age [334] above 30 years[335].

1850 - 2022

In the 100 years to 1950 life expectancy rose from 40 years to over 60 years due to advances in medical science, nutrition, sanitation, conditions of work and rising prosperity in Europe, North America, Oceania, Japan and parts of South America raising millions out of poverty and endemic disease[336]. In some countries such as Norway it was over 70 years. In Africa however life expectancy remained shockingly low. For people in Mali in 1950 it was 26 years while for Africa

[334] Generally considered to be from 3000 BC to 1200 BC.

[335] Marlowe F. *The Hadza: Hunter Gatherers of Tanzania* Berkeley: University of California Press cited Pinker S *Enlightenment Now* 2018 Viking.

[336] See at p 91 et seq below.

the average life expectancy was just 36 years. Someone in Oslo could then expect to live twice as long as someone in Kinshasa.

But the next 70 years began to close the gap. By 2019 average World life expectancy at 72.6 years exceeded that of any single country in 1950. Someone living in England born in the year of Churchill's speech can with some confidence expect to live well into their 80s. In Africa, even with far lower levels of industrialisation and prosperity, life expectancy has risen to over 60 years except for regional pockets of desert and drylands. -

In every part of the world people could now expect to live more than twice as long as their 19th century forebears.

Figure 32

Life expectancy, 2019

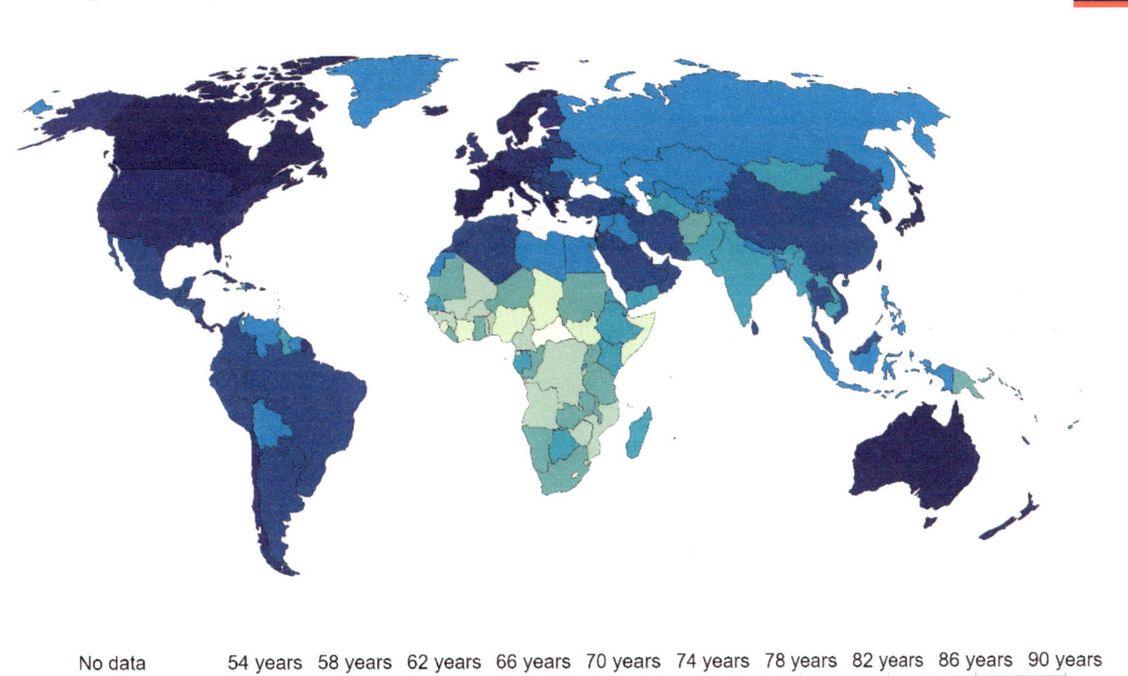

Source: Riley (2005), Clio Infra (2015), and UN Population Division (2019) OurWorldInData.org/life-expectancy • CC BY
Note: Shown is period life expectancy at birth, the average number of years a newborn would live if the pattern of mortality in the given year were to stay the same throughout its life.

The extending of life expectancy continues. The longevity of Queen Elizabeth 11 (96 years) would have staggered the Court of George III yet such a span it not uncommon today.

Nor is this trend confined to wealthy countries. In his survey of the golden age of the present[337] Johan Norberg describes how in Kenya in the ten years from 2003 to 2013 *"After having lived, loved and struggled for a whole decade, the average person in Kenya had not lost a single year of their remaining lifetime. Everyone got 10 years older yet death had not come a step closer".*

[337] *Progress: Ten Reasons to Look Forward to the Future* – 6 April 2017 Economist and Observer Book of the Year London Oneworld.

Infant and Maternal mortality

There have been precipitous declines in infant and maternal mortality in just 60 years.

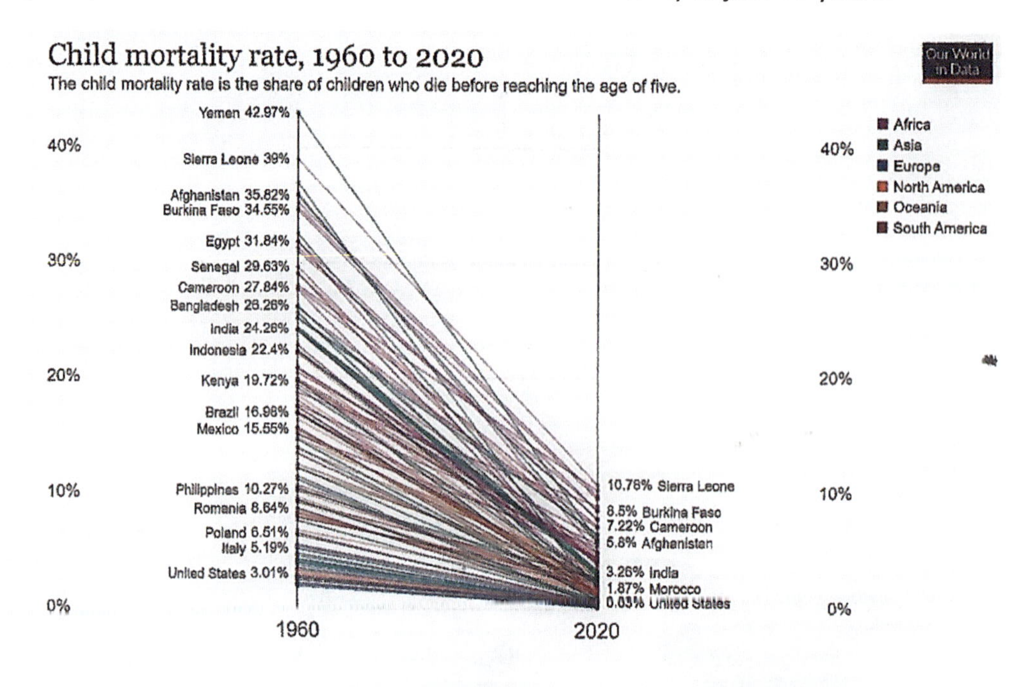

The risk of death in childbirth was almost sanctified by religion[338]. In the 19th century some 1,000 mothers and more died for each 100,000 births. Yet since 1860 there has been an overall steep decline of maternal mortality as is seen in the following chart.

Figure 34

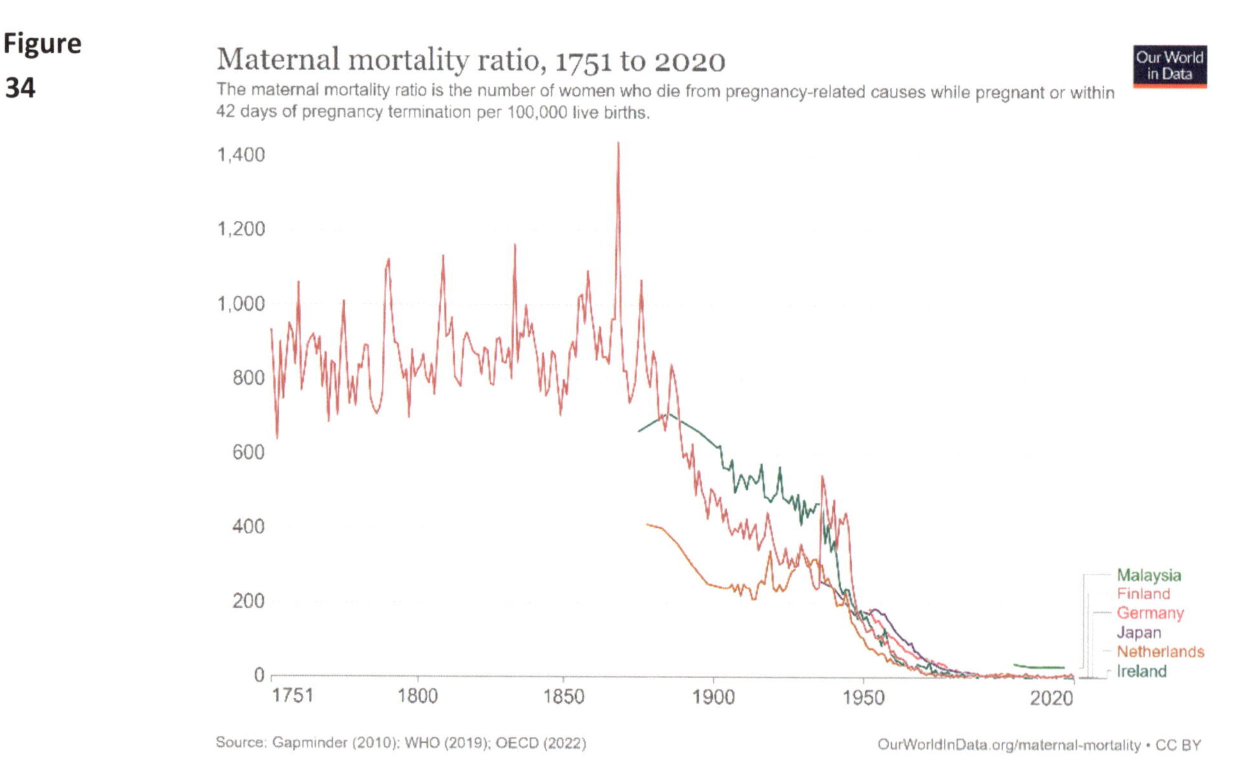

In Finland, which today has lowest risk of pregnancy related deaths, during much of the 19th century maternal mortality varied between 800 and 1,000 deaths for every 100,000 births. With

[338] Book of Genesis 3.16

each birth women faced a 0.9% chance of death. But with average births of 5 children such deaths were a common fact of existence.

The following chart also shows how maternal mortality in the United Kingdom has changed since then. Today, giving birth is 300 times safer than just a few generations ago.

Figure 35

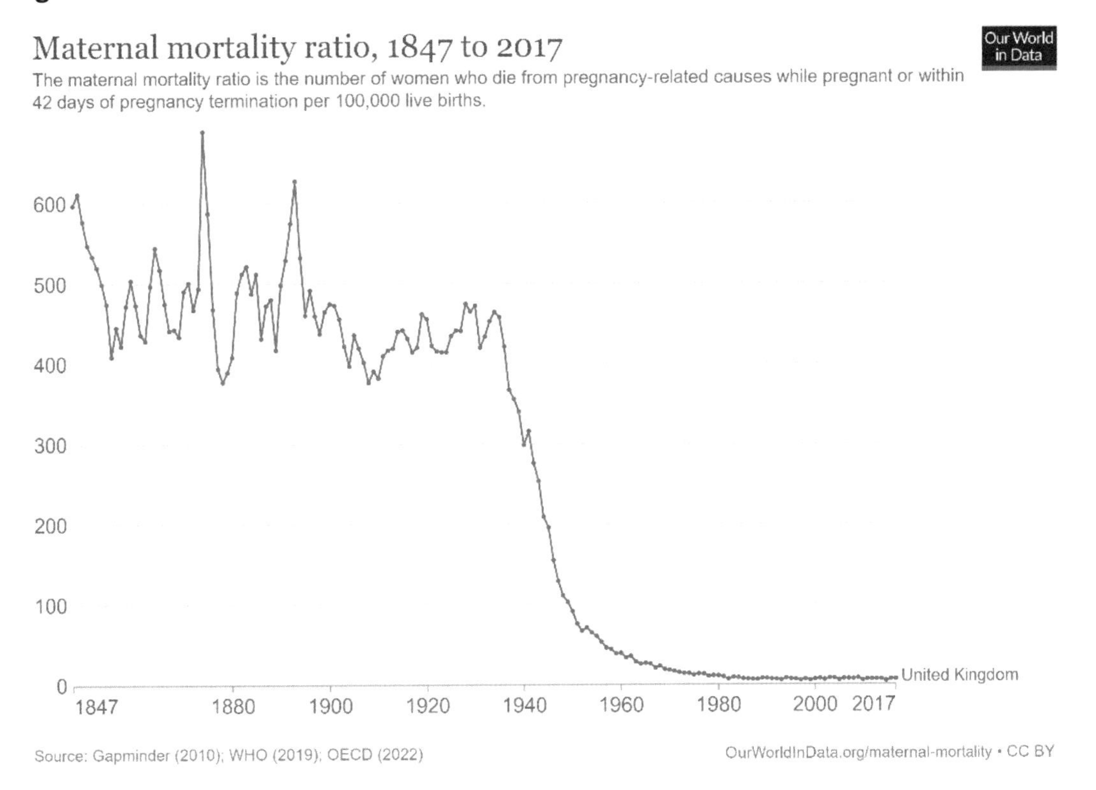

II. ELIMINATION OF DISEASE

Of all the blessings that scientific discovery has showered upon mankind none is greater than the dissipation of death by diseases which none could prevent or cure. The number spared from suffering and premature death runs into billions of human beings. Humanity having become accustomed to this priceless gift has little conscious remembrance of those who conferred it.

Just three examples illuminate how the gratitude of humanity is owed as much to those who spared it from death by disease as it is due to those who suffered death in war for its liberation.

Small pox

Smallpox is the only human disease[339] that has been successfully eradicated.[340] It was endemic across all continents. It was an infectious disease caused by the variola virus. It infected only humans and no treatment existed. It was fatal for millions of people.

---

[339] in May 1796, Edward Jenner inoculated a boy with cowpox, and then a few months later with the smallpox virus. When the boy did not develop any smallpox symptoms in response to being variolated, his hypothesis of the cowpox offering protection from smallpox was confirmed motivating his further research trials.

[340] Rinderpest has also been eradicated (2011) but was confined to animals – predominantly cattle and buffalo.

Figure 36 shows deaths in London due to smallpox as a percentages of all deaths from 1629 to 1902. In years of highest infection it inflicted death on one in four people.

Figure 36

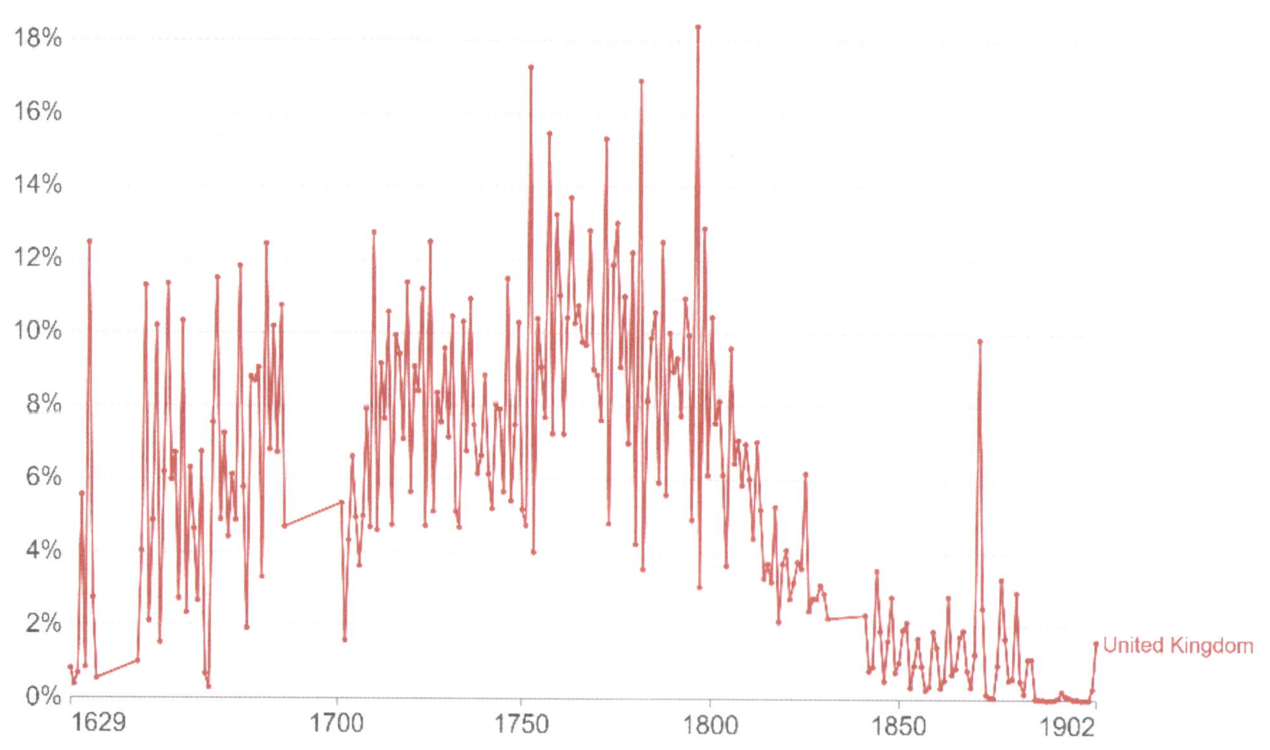

The scientific revolution also gave rise to the discovery of a vaccine to combat smallpox[341]. However vaccination was not adopted on a systemised universal scale until the early 1950s. It was the World Health Organization (WHO) in the aftermath of World War II that set standards for the production of smallpox vaccines.

Between 1920 and 1978 there were 11.6 million reported cases, though actual incidence was certainly higher. It was in the 1950s and 60s that there was a steep decline. With the exception of two tragic cases in UK due to a laboratory accident in 1978, the WHO could find no more. In 1980 it declared smallpox the first – and so far only– human disease to be eradicated globally[342].

A disease that was once endemic across the entire world which inflicted death upon countless millions and disfigurement of those who survived it had altogether vanished.

Figure 37

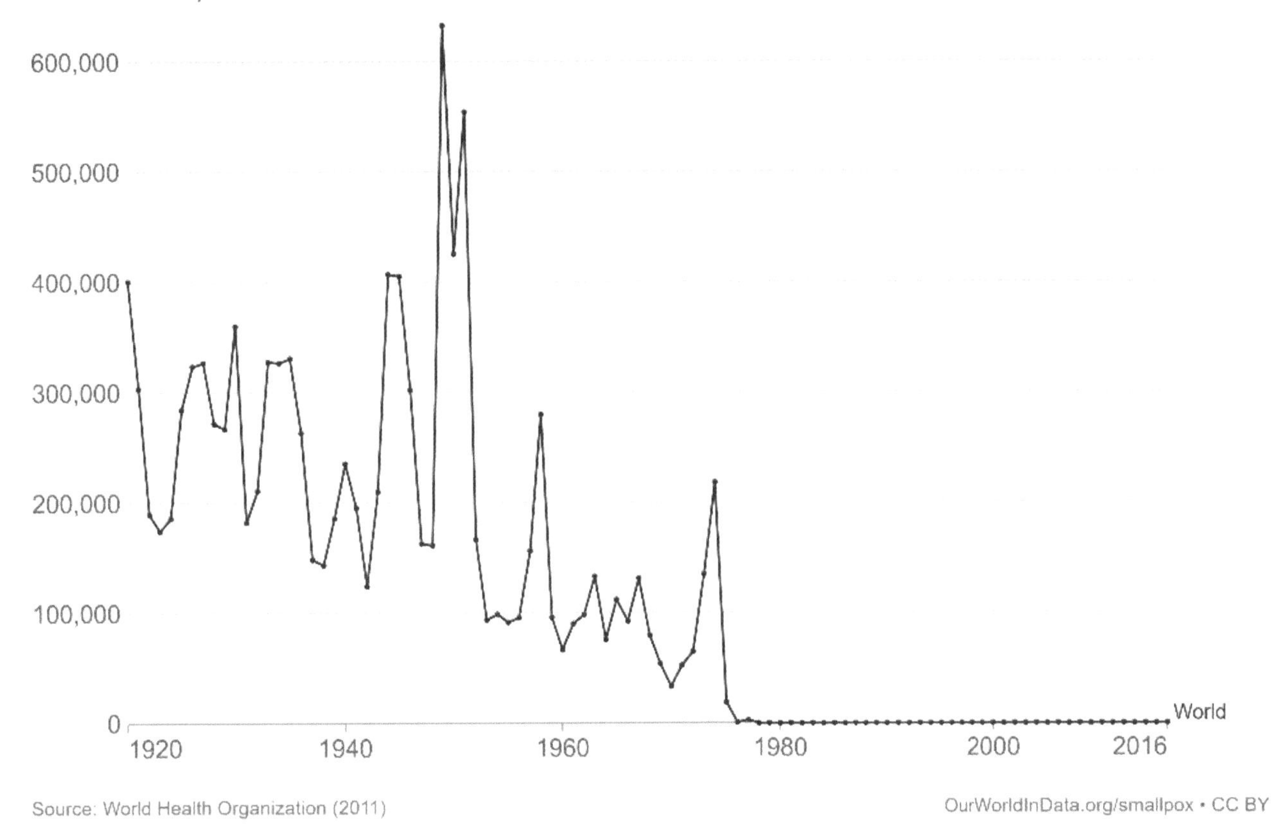

Poliomyelitis

Poliomyelitis (polio) is a highly infectious viral disease. It principally inflicts itself on children of under the age of five. The virus is transmitted by person-to person most commonly through the ingestion of infectious faecal contaminated material and spreads quickly in areas with poor hygiene and sanitation systems.

Humans are the only natural hosts for polioviruses. Once the gastrointestinal tract is infected, poliovirus may invade the central nervous system by penetrating the blood/brain barrier or by spreading along nerve fibres.

Two of three types of so-called 'wild' poliovirus have been eradicated but Type 1 (WPV1) causes paralysis in about 1 in 200 infections. There is no cure for polio. It can only be prevented by immunisation. It has existed for thousands of years[1] and has killed, maimed or paralysed millions. In the early 1980s there were still some 450,000 cases. Yet by 2021 there were only 6 confirmed cases, confined to just three countries[343].

[343] World Health Organisation . "*Global Wild Poliovirus 2016 - 2022*". Global Polio Eradication Initiative.

Figure 38

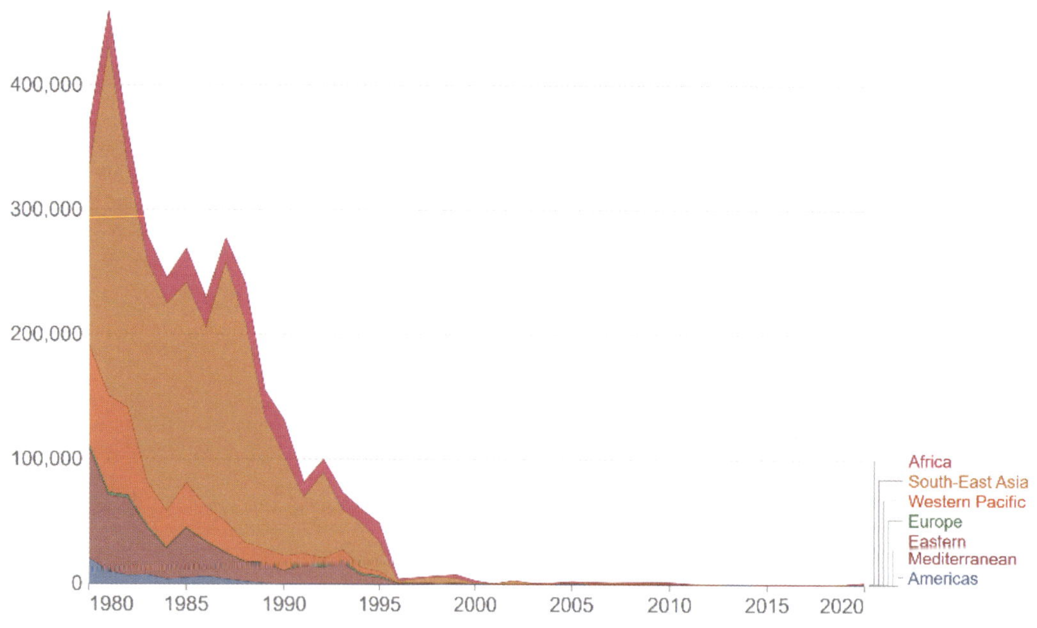

Source: Our World in Data based on World Health Organization and adapted from Tebbens et al. (2010) OurWorldInData.org/polio/ • CC BY

Salk vaccine

For this virtual elimination of a condition suffered by little children, that was terrifying by reason of its lifelong effects, humanity's gratitude is owed to Jonas Salk at the University of Pittsburgh. It was Salk who developed the inactivated poliovirus vaccine that bears his name. It was announced on 12 April 1955. Salt derived his vaccine from poliovirus grown in a type of monkey kidney tissue culture, which is chemically inactivated with formalin.

By the prick of a needle unimaginable horror was banished. After two doses of inactivated poliovirus vaccine, given by injection, 90% or more of individuals develop protective antibody to all three serotypes of poliovirus and at least 99% are immune to poliovirus following three doses.

Such an achievement is heroic. The City of New York the offered Salk the honour of a ticker-tape parade which his humility and modesty prevented him from accepting.

Malaria disease

Malaria is an acute illness caused by *Plasmodium* parasites. The bite of an infected female *Anopheles* mosquito transmits the parasite which lodges in the liver from the blood stream and reproduces. The parasite causes a high fever, shaking chills and pain. In severe cases malaria leads to coma and death. Of the 5 parasite species *P. falciparum* is the deadliest.

One of the abiding achievements of scientific discovery in the 20th century was its impact upon the peril of malaria mortality.

In 1900 malaria was endemic in half the world including in the United States. It is estimated to have killed 2 million people each year. Archaeological examination of teeth and bones has revealed that it was rife in the marshy valleys of the Tiber during the Roman empire and enfeebled defence of its capital.[344] Oliver Cromwell[345] succumbed to it. As late as 1965 outbreaks occurred in Europe, Russia, Australia and the United States.

Since then it has been eliminated in more than 100 countries - all but 27% of the land surface of the Earth. By 1999 the global malaria mortality had declined by 90%[346]. Between 2000 and 2015 world malaria death rates fell by a further 60% averting 6.2 million deaths. It declined most steeply among young children in Africa where it fell by 71%[347].

Treatment of malaria with quinine[348] in the 19th century was the first known use of a chemical compound to treat an infectious disease. Quinine ameliorated the affliction but the major advance in both control and also of treatment of the disease was made in the second half of the 20th century.

Malaria Control

Whilst the drainage of swampland for conversion to agricultural land restricted breeding grounds for mosquitoes, pesticides are the main weapon in the war on malaria. DDT[349] was the principal insecticide until the 1970s. It is one of twelve WHO approved insecticides[350].

Following the reaction to the sentimentalised gloom of *Silent Spring* the use of DDT was banned in many countries as an agricultural insecticide due to its perceived impacts in the food chain. However it remains extremely effective for control of mosquito borne malaria in many countries including India and African states where it is at its most prevalent.

Alternative insecticides have been developed including DFDT a related compound of DDT[351]. Pyrethroid insecticides are also widely used alternatives to DDT. They were developed in the 1970s by the oldest agricultural institute in the world at Harpenden in the UK[352] benefitting from studies by other scientists in the 1920s[353]. The pyrethroid insecticides were a major advance having low toxicity for mammals and fast biodegradation.

The twin uses of insecticide for control of malaria are the spraying of the inside of dwellings and the impregnating of bed netting. Spraying is used on walls made from porous materials including mud or wood. In 2008, 44 countries employed spraying of dwellings for malaria control. Insecticide-treated nets (ITNs) used over beds confer over 70% protection and are twice as effective as untreated nets.[354] Between 2000 and 2008, the use of ITNs saved the lives of an

---

[344] Sallares R. (2001) – *Malaria and Rome: A History of Malaria in Ancient Italy* World in Data.
[345] The Cromwell Association *"Cromwell's Health and Death"* https://www.olivercromwell.org/wordpress/about-us.
[346] Box 4.1 Malaria-related mortality in the 20th century" in the World Health Organization's World Health Report (1999). World in Data.
[347] World Health Organisation *World Malaria Report* 2015 Geneva cited in *Progress* Norberg J.2017.
[348] Quinine was first isolated in 1820 from the bark of a cinchona tree of Peru and its molecular formula was determined in 1854.
[349] Dichlorodiphenyltrichloroethane. Its insecticidal action was discovered by the Paul Müller in 1939. In World War II it was used to limit the spread of malaria and typhus among civilians and troops. Müller was awarded the Nobel Prize in Physiology or Medicine in 1948 .
[350] Sadasivaiah S, et al (2007). *American Journal of Tropical Medicine and Hygiene*. 77 (6 Suppl):pp 249–263.
[351] difluorodiphenyltrichloroethane (DFDT).
[352] Rothamsted Research institute founded in 1843.
[353] Hermann Staudinger and Leopold Ružička. Each were separately awarded the Nobel Prize for chemistry.
[354] Raghavendra K, et al (2011). "*Malaria vector control: From past to future*". Parasitology Research. 108 (4): 757–79.

estimated 250,000 infants in Sub-Saharan Africa.[355] By 2015 68% of African children were using them.[356]

Treatments of malaria were made much more effective with the discovery in 1972 of Artemisinin[357] Artemisinin-based combination therapies are now the essential treatment worldwide for the most dangerous strain *P. falciparum*. They kill all the life cycle stages of the parasites and act faster than other treatments.

It now appears that an effective vaccine may also become available. Researchers from the University of Oxford reported findings in 2021 from a trial of a candidate malaria vaccine with efficacy of 77%[358]. It is the first to meet the WHO goal of at least 75% efficacy.

Measles, Hepatitis, Influenza

The depth and extent of the beneficial advance of the health of the human race in its conquest of infectious disease over the last 70 years is symbolised and was largely the result of the work of Maurice Hilleman.[359]

one of eight children he was brought up on his uncle's farm on the death in childbirth of his mother. It is said that he applied for the job of manager of a store and was rejected on the grounds of an unsuitable personality for retail. His forcefulness later became legendary among his laboratory staff who were both fearful of and also devoted to him. Hilleman's gift to humanity was his development of no less than 25 vaccines. These comprehend most of those which are today recommended for children.

He first developed a vaccine for Japanese B encephalitis for American GIs in the Pacific during World War II. Shortly after that war he discovered the impact of genetic change on mutation of the influenza virus allowing him to advance the need for yearly vaccinations. His vaccine enabled the world to combat the Asian flu epidemic of 1957 saving millions of lives[360].

Hilleman's extraordinary achievements included creation of the mumps vaccine and the Hepatitis B vaccine. It is the strain used in the trivalent MMR vaccine for measles, mumps and rubella - the first multiple virus strain vaccine to be approved. His work on the hepatitis vaccine resulted in its use in 150 countries and in the collapse in incidence of the disease. Its benefit to transplant surgery is impossible to exaggerate.[361]

Hilleman defeated measles. The disease is extremely contagious: nine out of ten people who are not immune and share living space with an infected person will be infected. In dense populations with poor nutrition and healthcare, fatality rates have been as high as 28%.[362]] In sufferers with

[355] Howitt P, Darzi A et al , *(2012). "Technologies for global health"*. The Lancet. 380 (9840): 507–35.
[356] *'Achieving the malaria MDG target: reversing the incidence of malaria 2000–2015'). UNICEF. WHO. September 2015..*
[357] Discovered by Tu Youyou, who shared the 2015 Nobel Prize in Physiology or Medicine for her work .Artemisinin is extracted from the plant *Artemisia annua*, sweet wormwood, a herb employed in Chinese traditional medicine.
[358] Malaria vaccine becomes first to achieve WHO-specified 75% efficacy goal, News Release 23 April 2021, University of Oxford.
[359] Microbiologist and vaccinologist 1919 – 2005.
[360] Myron M. Levine & Robert C. Gallo (2005) A Tribute to Maurice Ralph Hilleman, Human Vaccines, 1:3, 93-94, DOI: 10.4161/hv.1.3.196.7
[361] *"Controlling the hepatitis B virus scourge ranks as one of the most outstanding contributions to human health of the twentieth century...Maurice removed one of the most important obstacles in the field of organ transplantation"* Thomas Starzl liver transplant pioneer. See Paul A. Offit *'Vaccinated : one man's quest to defeat the world's deadliest diseases'* 2008, ©2007 Smithsonian Books; Collins.
[362] Perry RT et al (2004). *"The clinical significance of measles: a review"*. Journal of Infectious Diseases. 189 Suppl 1 (S1): S4-16. Wikipedia.

other conditions (for example AIDS) the fatality rate is even higher. Since the mid-19th century measles is estimated to have killed 200 million people.

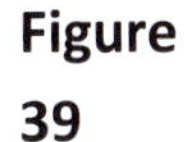
Figure 39

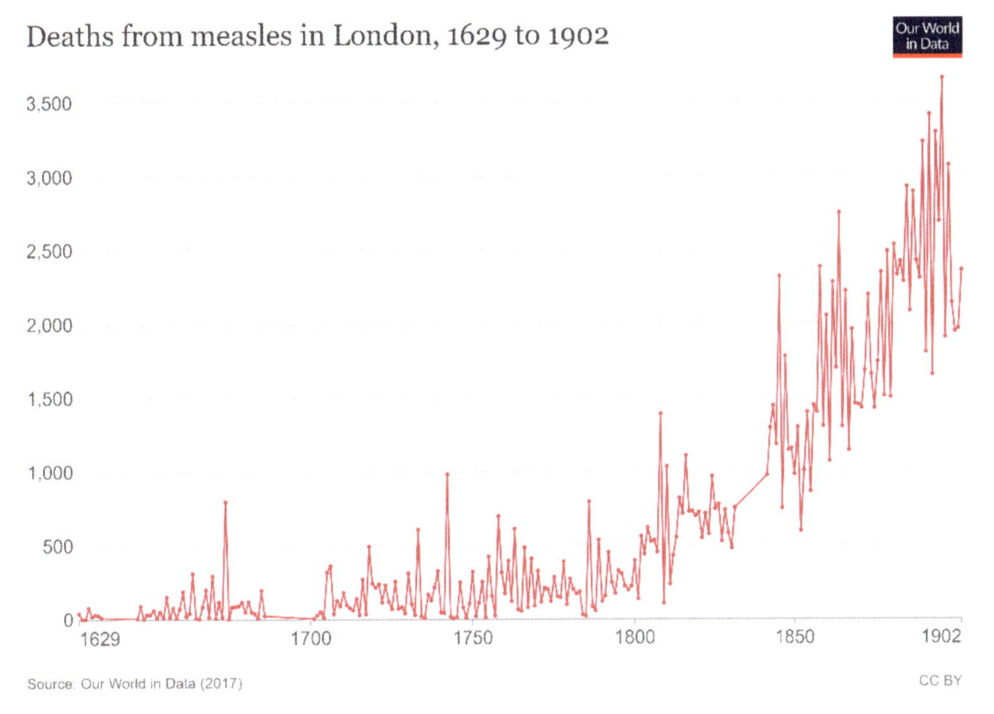

The Hilleman measles vaccine resulted in an 80% decrease in deaths between 2000 and 2017. Measles is still one of the most vaccine-preventable disease causes of death.[363] With increasing acceptance of vaccination the death toll had fallen to 73,000 by 2014 - down by 95% since 1980.

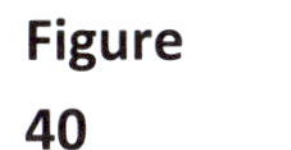
Figure 40

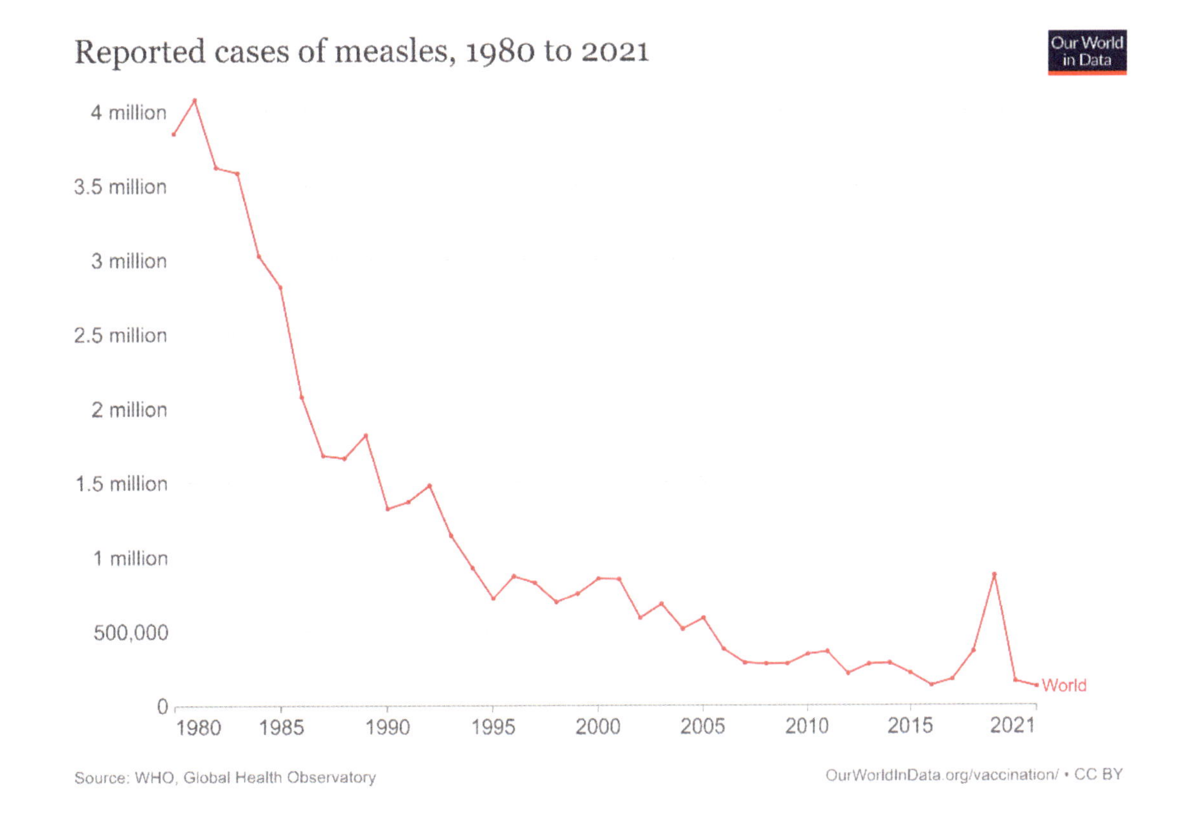

[363] Russell, S.J. et al (2019)"*'Oncolytic Measles Virotherapy and Opposition to Measles Vaccination*". Mayo Clinic Proceedings. 94 (9): 1834–39. Wikipedia WHO records show that In 1980 no less than 2.6 million people died from disease.

III. FOOD

As has been seen (Part 4) the predictions of famine and collapse of society that would arise with the multiplication of the population were falsified by reality. From Malthus to Vogt to Erlich and the Club of Rome and more recently Attenborough[364] such prophesies have been confounded.

What undermines all such predictions is the unpredictability of the future - as witness the steep declines in replacement rates of births. Nor can a useful forecast be made as to what may be the fruit of the inherent qualities of mankind for spontaneous evolution and adaptation to change in conditions of freedom and opportunity.

In countries dominated by totalitarian central planning famine has prevailed[365]. Forced collectivisation and confiscation precipitated famine in Ukraine with Stalin's enforced exile or murderous elimination of the Kulak farmers, in the Maoist Great Leap Forward in China, in Pol Pot's Year Zero and most recently in Kim Jong's Arduous March of the last years of the 20th century. When Maoist communes reverted to individual farm ownership the nightmare of famine in China[366] vanished. In just 20 years grain yields increased by 600%. It created food surplus available for world markets[367].

There are countless examples of the immense benefits of industrialisation for the feeding of masses of population. Simply consider the convulsive revolution just in transport: the canals or the bulk grain trucks of the United States or the refrigerated transport of lamb from New Zealand and Australia[368]. All provided cheap and abundant food of high nutritional value for the great masses of humanity.

However of all the advances in nutrition it is the chemical revolution that provided the most enduring and extensive consequences for the feeding of mankind.

Fertiliser from the air

In the 5 years before World War 1 Carl Bosch found the means to industrialise the laboratory method discovered by Fritz Haber for fixing nitrogen. By the use of large compressors and high pressure furnaces a process was devised for producing synthetic nitrate. Each of these stages was hazardous and complex. A primary commodity for the process was pure hydrogen gas. The process also yielded the product ammonia. The outcome was that it became possible to produce ammonia synthesised into nitrate on a massive scale.

Is so doing Bosch brought about an immense intensification of food production. He had effectively taken a gas and made it a fertiliser. It mirrors the miracle of photosynthesis whereby carbon dioxide in the atmosphere, light, water and the enzyme rubisco creates the food of plants

---

[364] In a 2013 interview with the Radio Times, Attenborough described humans as a "plague on the Earth "and described the act of sending food to famine-stricken countries as "barmy" for population reasons. He warned 'human beings have overrun the world' in new film. *inews.co.uk*. 15 January 2020.

[365] Devereux S. 2000 *Famine in the Twentieth Century*. Institute of Development Studies UK cited in Pinker Op cit p 78.

[366] Dikötter, Frank. *Mao's Great Famine: The History of China's Most Devastating Catastrophe, 1958–62*. Walker & Company, 2010. p. xii ("at least 45 million people died unnecessarily").

[367] Kate Xiao Zhou *How the Farmers Changed China: Power of the People*. Boulder. Westview Press 1996 cited in Norberg J. "Progress" Oneworld Publications 2017.

[368] The author's great grandfather founded the Shaw Savill shipping line for the New Zealand and Australian trade which prospered by shipping of refrigerated meat.

and thus of mankind. For his remarkable achievement Bosch was awarded the Nobel prize for Chemistry in 1931.

The Haber–Bosch process produces 100 million tons of nitrogen fertilizer every year. It is responsible for feeding roughly one-third of the human population.[369] It has been well observed that if the food grown today had to be grown with pre-nitrogen techniques of farming it would require that an area the size of Russia to go under the plough,[370]

The Green Revolution

The chance encounter between Vogt and Norman Borlaug in Mexico (see Part 5) symbolises the essential distinction between the static and the dynamic conceptions of resources and wealth – between the fixed cake or 'lump' theory of Malthus and Erlich and the organic yeast of discovery and invention that is the reality.

The greatest contradiction of Malthus of the mid-20th century was without question the work of Norman Borlaug – Vogt's chance companion. Engaged by the Rockefeller Foundation to inculcate adoption of new methods of agricultural development by Mexican farmers he embarked on fastidious and extensive trials of cross breeding of wheat types.

Wheat has four times as many genes as humans. This provides a store of potential variations but it also renders discovery a task of intimidating complexity. Borlaug ultimately succeeded, with repeated cross breeding, in growing rust resistant[371] and photoperiod[372] insensitive varieties. He was then able to cross breed these with very high yield wheat types so producing five new varieties which were in addition highly productive.

He also devised a solution to a grave problem of his high yields – that of 'lodging'. Borlaug's new varieties required chemical fertiliser to supplement the poor Mexican soil and these were found to bear huge heads of grain. The consequence was that in high winds or severe rain the plants, being top heavy, fell over, thus causing the entire crop to succumb to a general collapse. The bending of the stems prevented water and nutrients from supplying the juvenile grain.

Borlaug undertook the arduous task of breeding short stem wheat types. The development of short straw wheat stem varieties was complex due to the natural resistance to reversal of the evolution of long stems which had more prospect of sun exposure and thus more likely to grow and reproduce.

Borlaug's achievement in creating short stem highly productive wheat was enhanced by the fact that less energy was expended by the plant growing inedible stalks. When introduced all over Mexico in 1963 the harvest was six times that in 1944 when he had first arrived there.

---

[369] Vaclav S. (2001). *Enriching the earth*. Cambridge, Massachusetts: MIT Press.

[370] Morton O.The Planet Remade? How Geoengineering Could Change the World Princeton University Press 2015 p 204.

[371] Wheat leaf rust (*Puccinia triticina*) is a fungal disease that affects wheat, barley, rye stems, leaves and grains. In temperate zones it is destructive on winter wheat because the pathogen overwinters. It severely reducing wheat crop yields.

[372]Photoperiodicity is the phenomenon of ssensitivity to the length of a day and it's that which controls wheat growth of winter wheat, It operates as biochemical clock that measures day length. By genetic variation it proved possible to render wheat insensitive to photoperiodicity. Thus wheat would sprout and grow without deferral.

That same year (1963) Borlaug moved to India and Pakistan each of which faced mass starvation. Despite the war between those countries Borlaug procured the shipment of thirty five trucks of high yield grain from Mexico. Even with late planting crop yields rose by an astonishing 70%. Impelled by the imminence of starvation both governments authorised a wide planting programme resulting in a vast increase in yields.

Figure 41

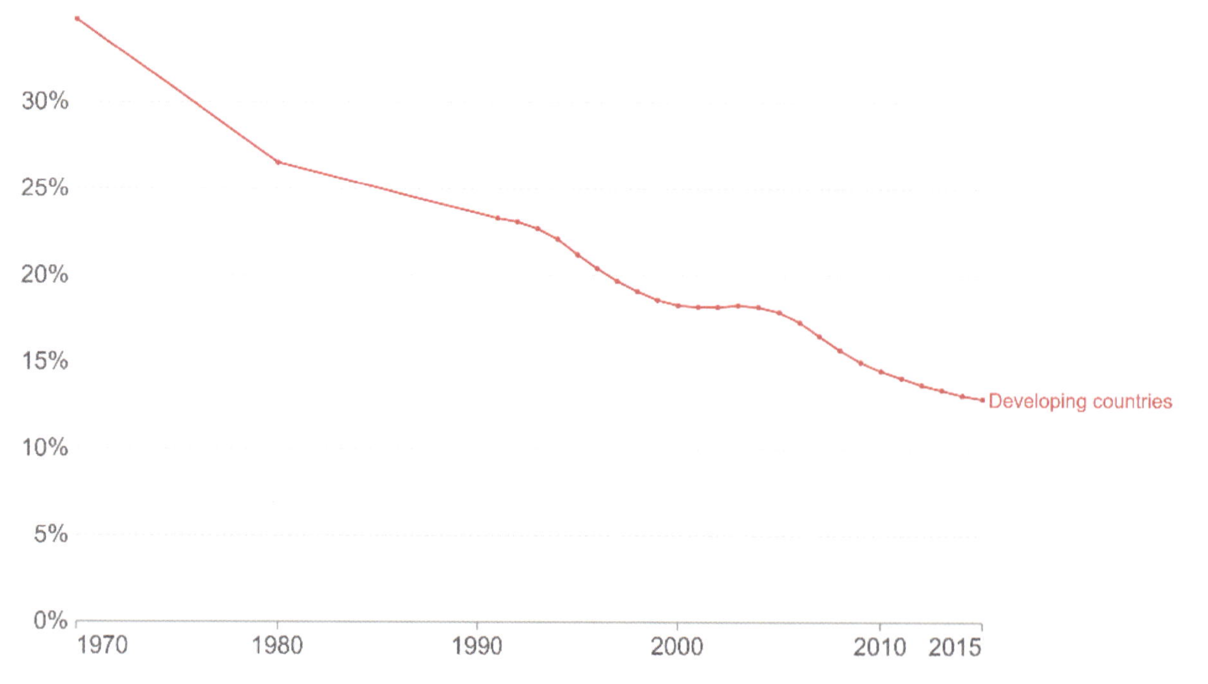

Farmland increased by a mere 12% while global farm production rose by 300% in the 48 years to 2009. Borlaug was *"The first person in history to save a billion human lives"*[373]

The modest rise in CO_2 over the past 60 years to the rise in crop yields has been remarkable both in its extent and in correlation. Set out below are Figure 42 graphs of rises in CO2 emissions and atmospheric density and Figure 43 graphs of long term wheat yields in Europe from 1850 to 2015. Crop yield increases of over 100% have occurred since 1960 in almost exact correspondence with rise in atmospheric CO_2.

Figure 42

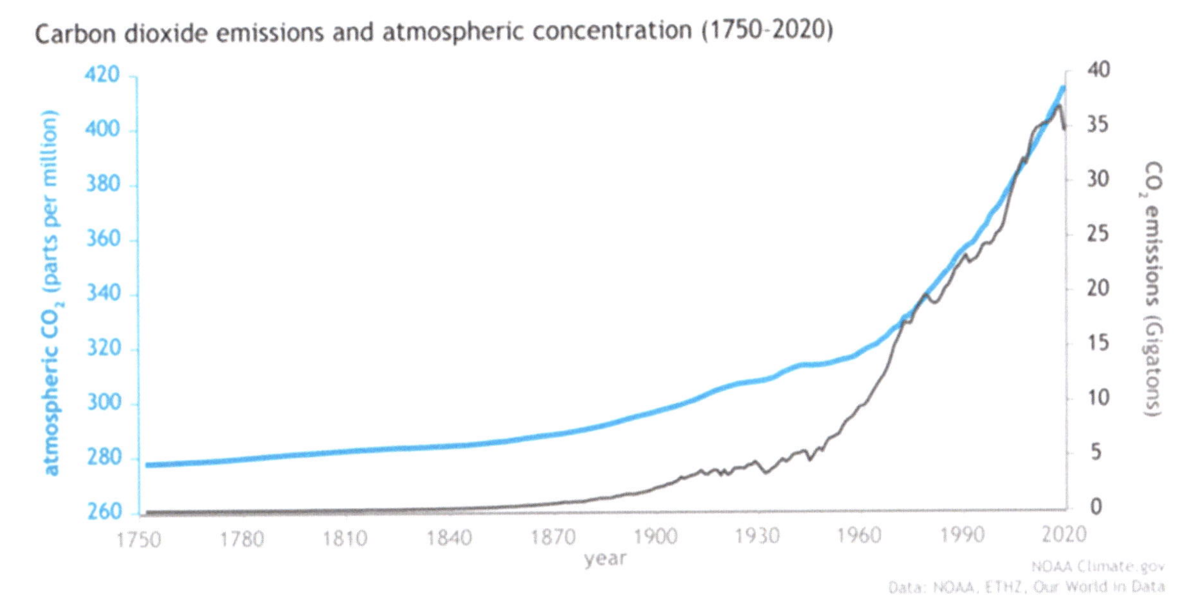

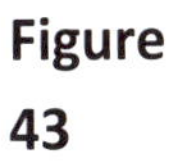
Figure 43

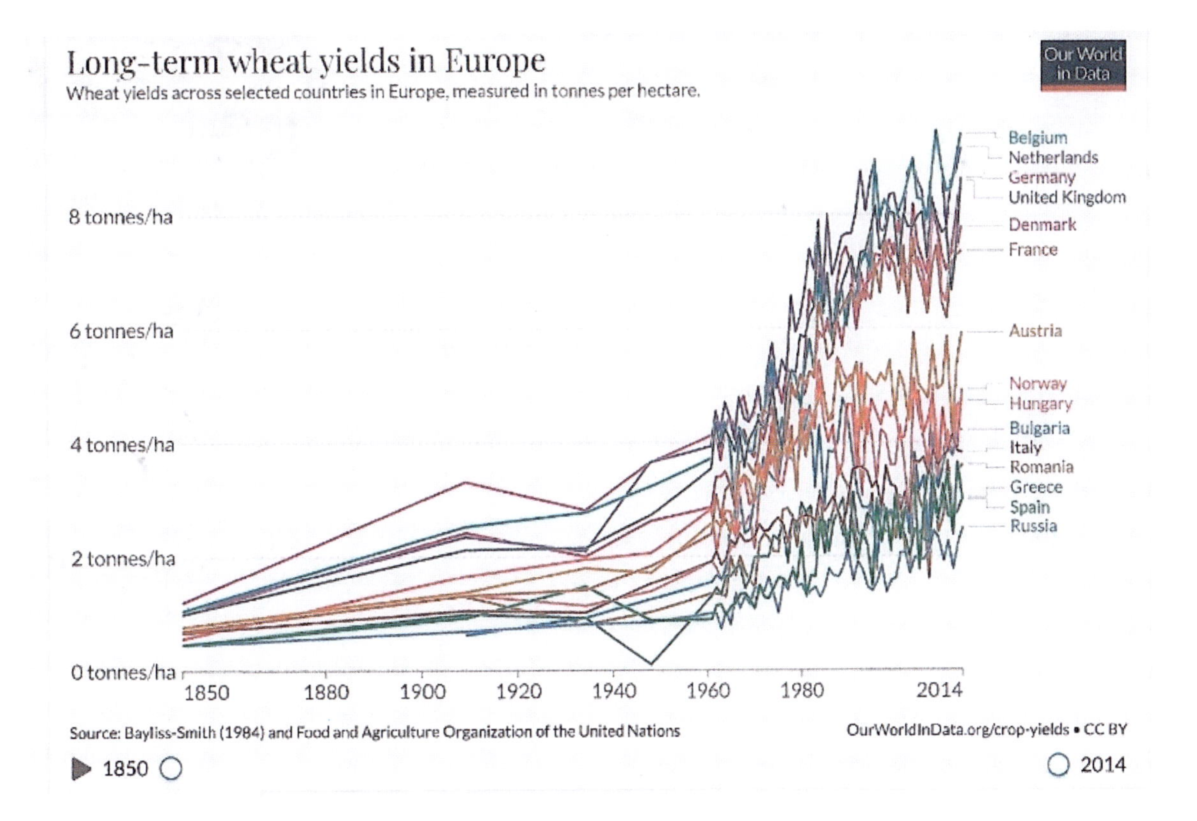

[373] Senator Rudy Boschwitz See Preevor J. *"Feeding the World in 2050"* American Food and Ag Exporter 2009 10-14 at p.14 In 1970 Borlaug was awarded the Nobel Prize for increasing global food supply.

IV. ABOLITION OF EXTREME POVERTY

Wherever economic freedom has prevailed there has been an eruption of prosperity and collapse of extreme poverty. The notion of the inevitability of poverty died with the emergence of open market economies. Since 1820 World GDP has multiplied by a factor of over 100.

Figure 44

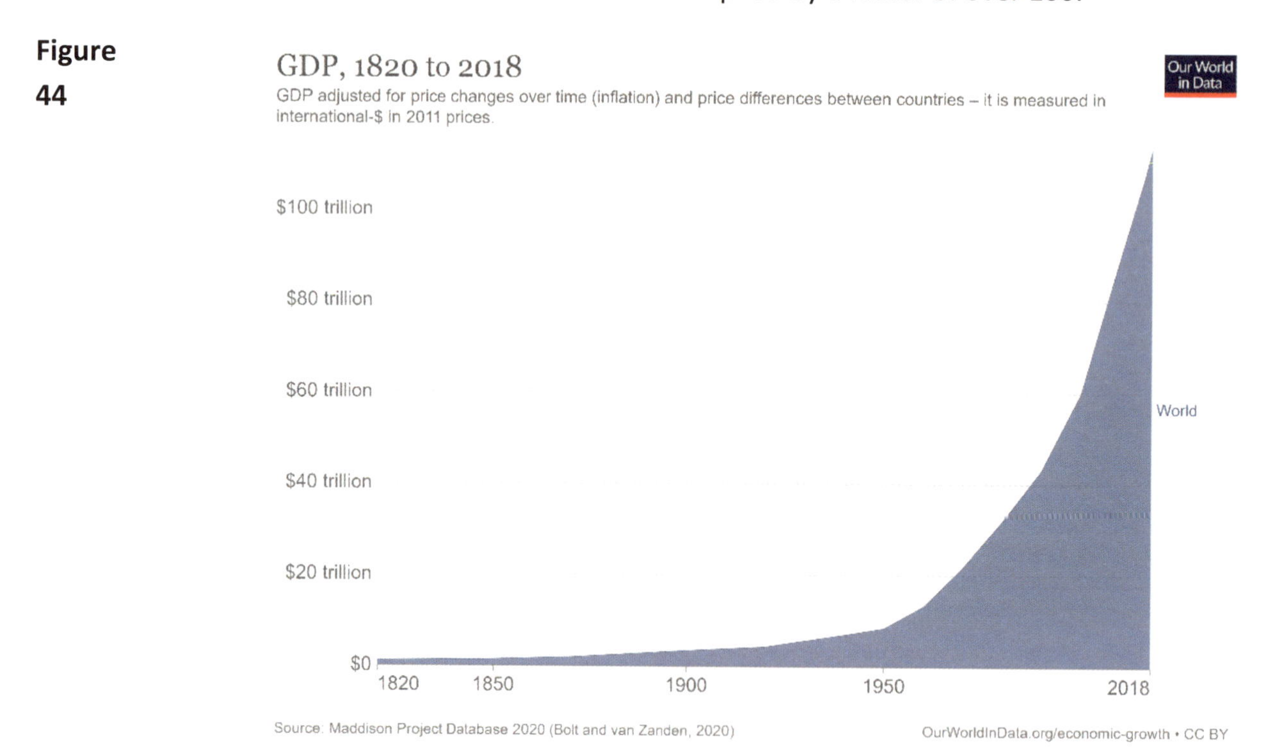

Economic inequality before 1950 was shockingly high. In 1820 94% of global population lived in extreme poverty[374]. The destitution only changed with the economic growth in free markets.

Figure 45

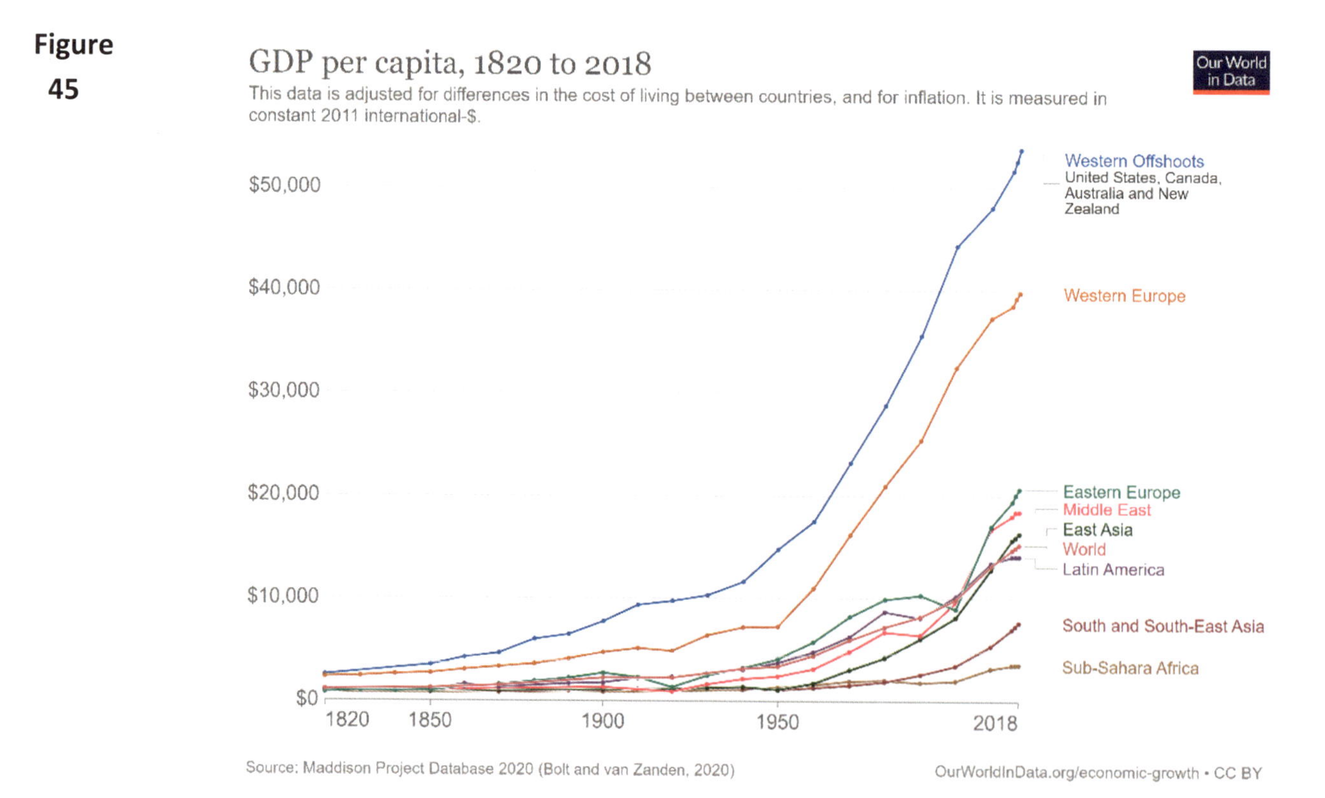

[374] Bourgignon F. et al *'Inequality among world citizens 1820 – 1992'* American Economic Review 92, 4 2002 pp727-744.

The ascent[375] from extreme poverty had been made in most western countries by 1960. It was then eradicated in Japan, South Korea, Taiwan, Hong Kong and Singapore with steep falls in China (1979) and India (1991) with their integration in the global economy.

Figure 46

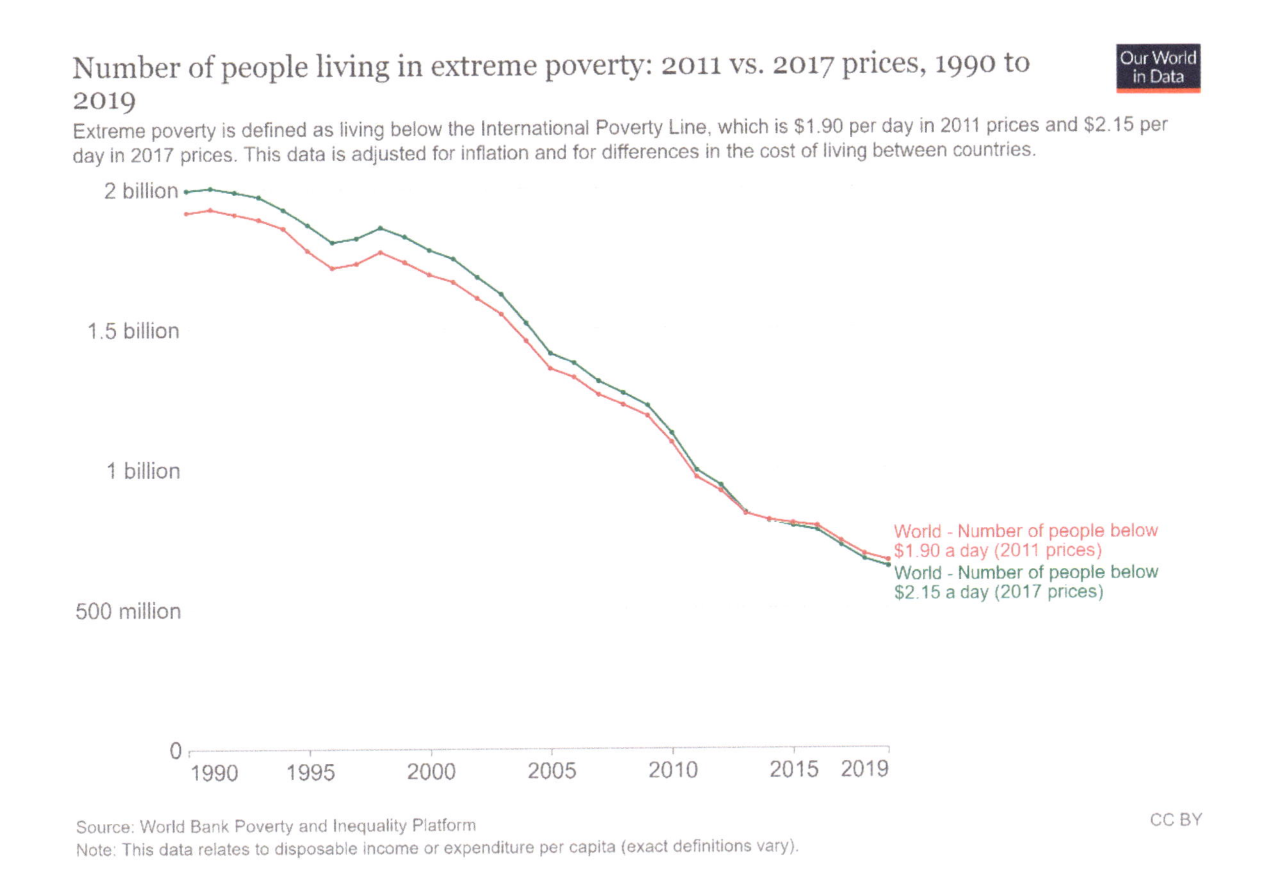

The truly epic rise in prosperity and levels of subsistence was accompanied by a vast increase in population. From 1990 to 2015 world population grew by more than 2 billion yet number of people living in extreme poverty fell by more than 1.5 billion.

The transformation occurred as **labour intensive** production began to be superseded by **technology based** production. With further growth – but not depletion of resources –came **knowledge intensive** production in services: finance, research, intellectual property and consultancy services of all kinds. Open market economies at different stages of development make people richer in one place thus making other people in other places richer at the same time.

In conclusion there are set out below diagrams depicting the advances in safety, security, tolerance, freedom and opportunity for human development and prosperity that have transformed human existence over a few generations.

They are their own proof.

[375] World Bank Development Report 1997 *'The State in a Changing World'* Washington DC.

Figure 47

Major world conflicts over more than 500 years shown by their duration and the extent of peace.

Figure 48

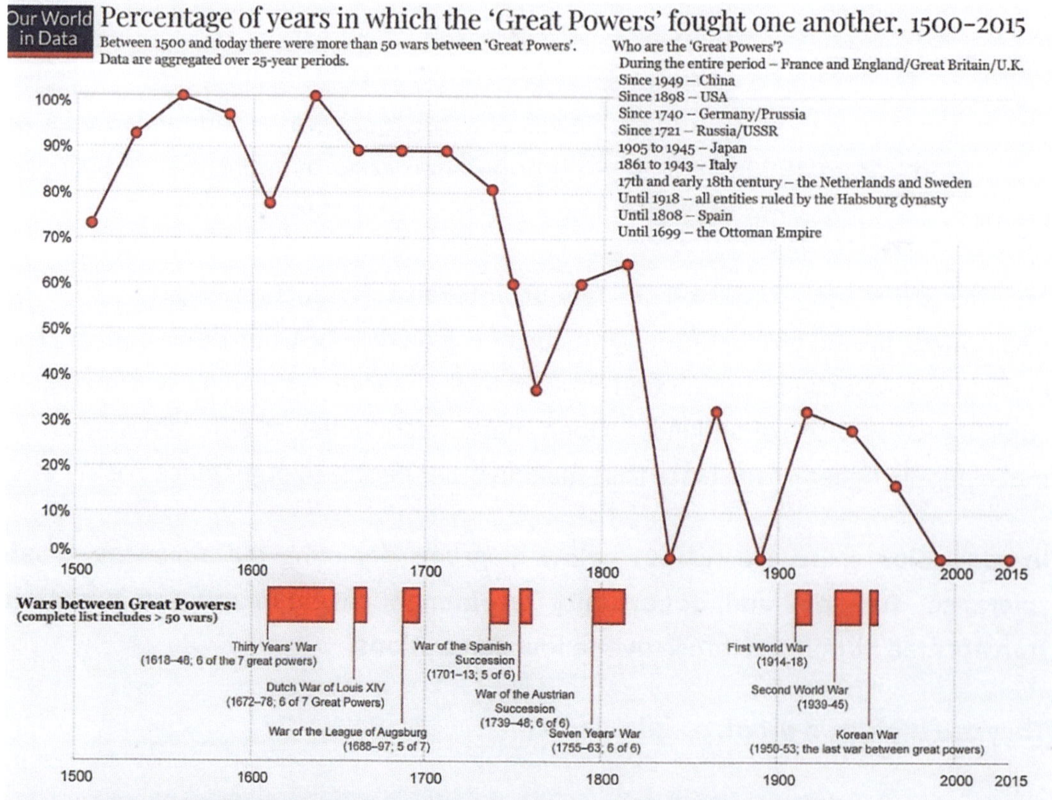

VI. TOLERANCE & FREEDOM

Figure 49

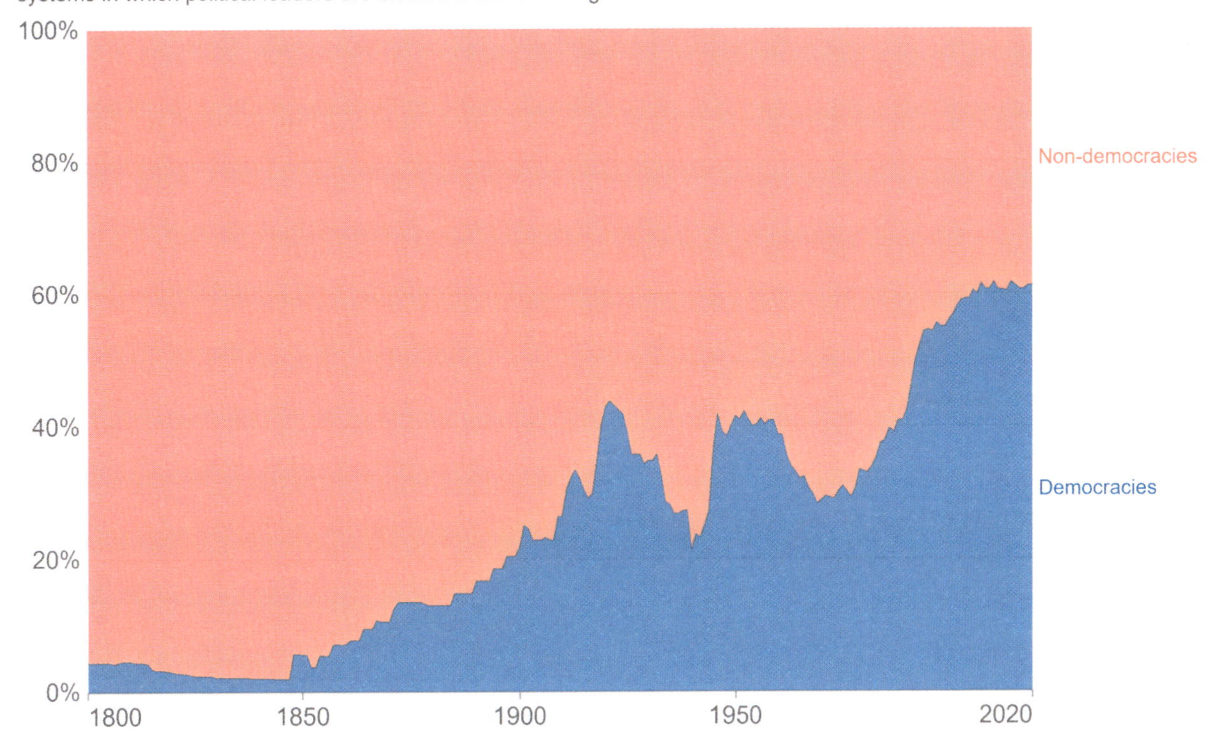

Figure 50

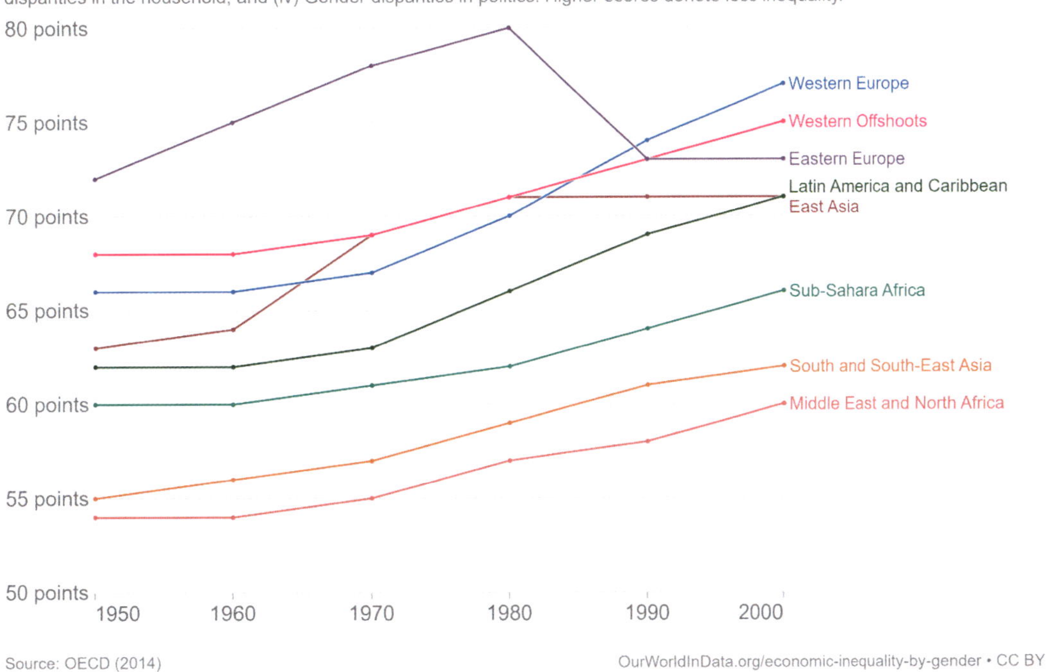

Figure 51

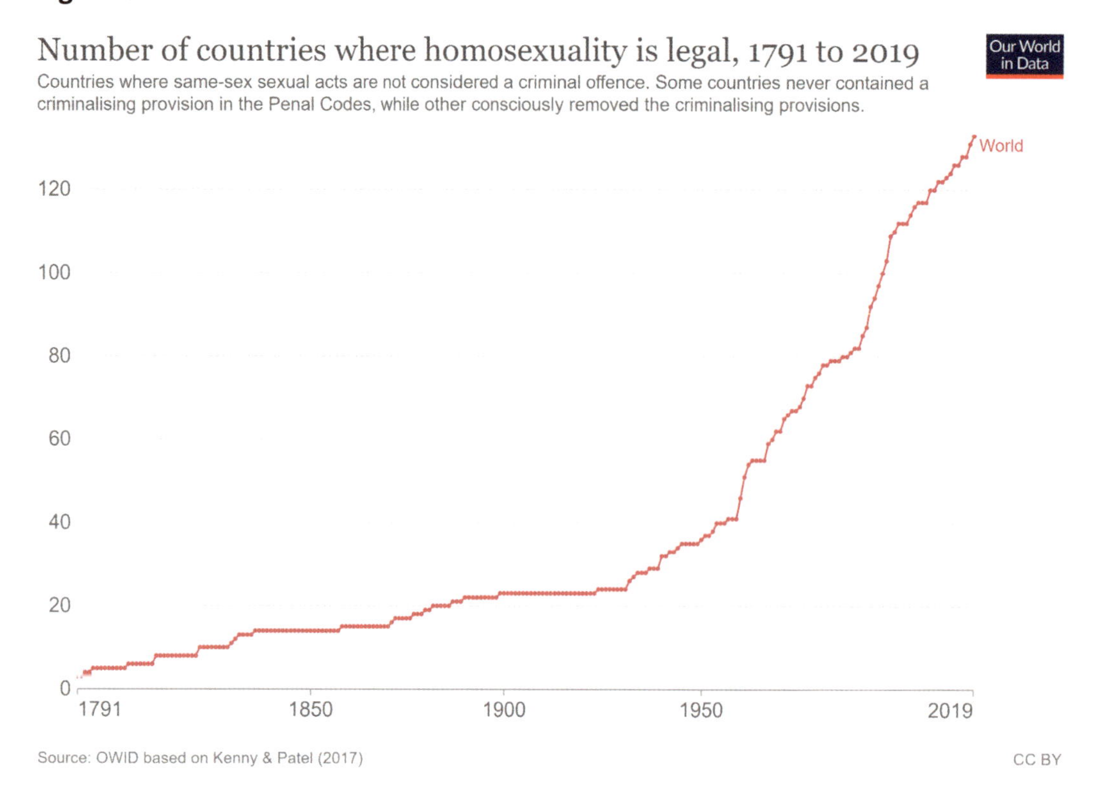

VII. HUMAN DEVELOPMENT

The Human Development Index is published by the United Nations Development Programme and this data is shown in the time-series chart here.

Figure 52

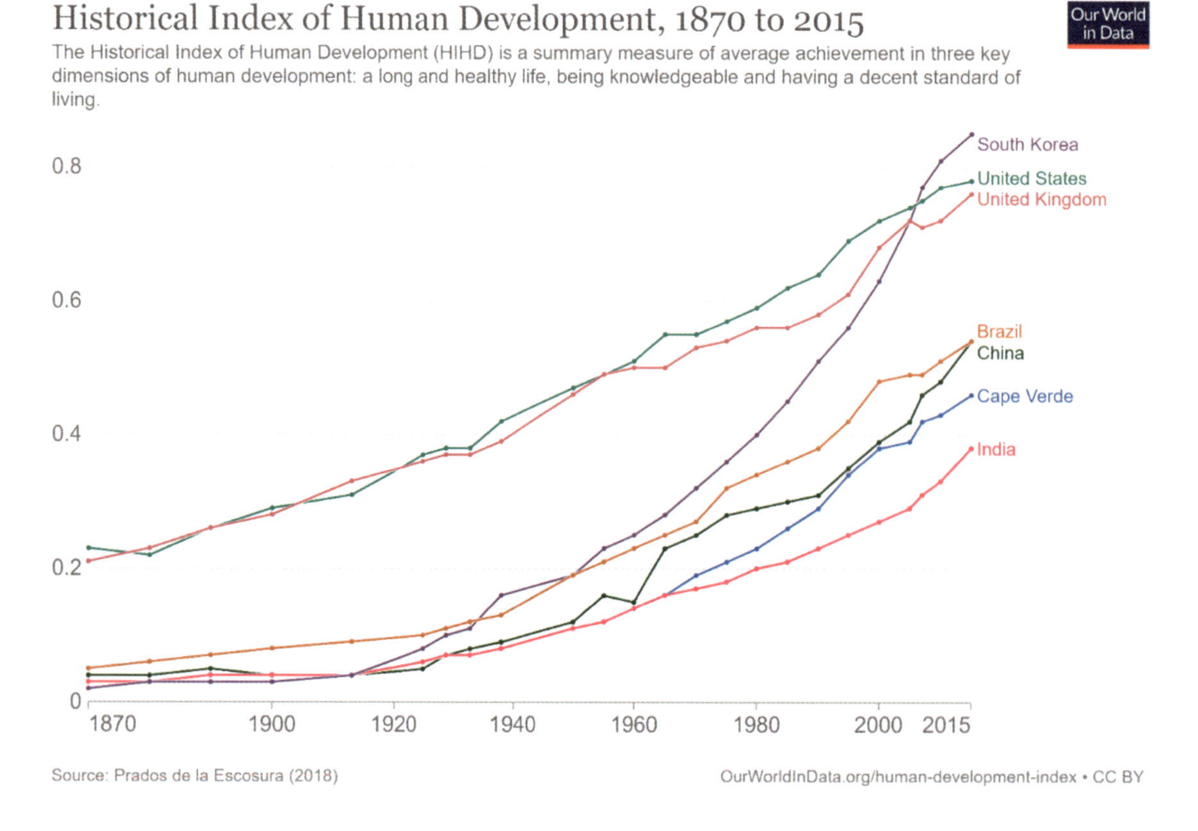

Figure 53

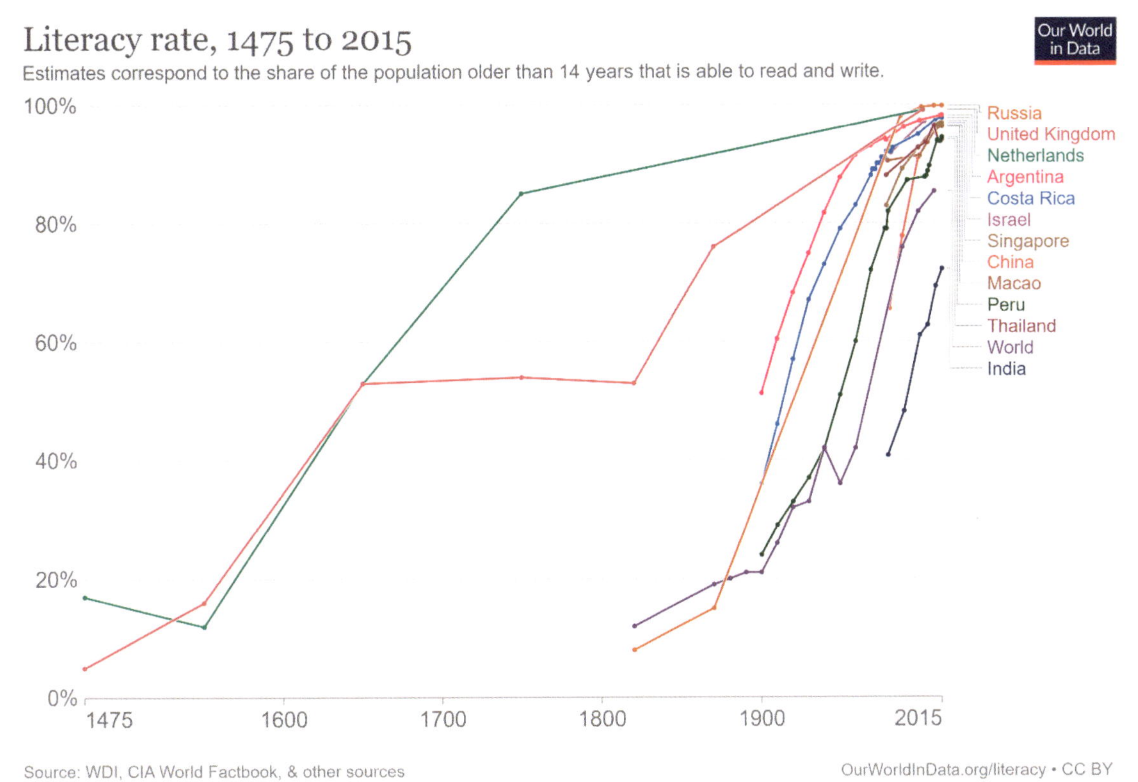

Conclusion

Greater than the blessings of material wellbeing that have been showered on humanity by the industrial, scientific and economic revolutions is the sense of self-worth, dignity and security they have implanted in billions of human beings.

By bringing near the defeat of extreme poverty in all but those countries torn by war, civil strife or under tyranny, a mortal blow has been inflicted on its horrors. For billions of people there has arisen a radiant hope for an end to

> *'Loss, grief, anguish, worry, over-thinking, madness, frustration, anger, alienation, humiliation, shame, loneliness, depression, anxiety and fear'*[376]

Surely we now live upon the broad and sunlit uplands.

SUMMARY

On every measure of wellbeing -- life expectancy: elimination of disease: abundance of food: defeat of poverty: safety: security: tolerance: freedom: prosperity and human development - liberal democracies and open markets have within just four generations so transformed human existence as to inspire wonder and gratitude to a merciful and generous Providence.

[376] World Bank *Voices of the Poor* 2016 Cited by Norberg J. Op cit. Based on interviews with 60,000 men and women from over 60 countries.

The Bruges Group is an independent all-party think tank. Set up in 1989, its founding purpose was to resist the encroachments of the European Union on our democratic self-government. The Bruges Group spearheaded the intellectual battle to win a vote to leave the European Union and against the emergence of a centralised EU state. With personal freedom at its core, its formation was inspired by the speech of Margaret Thatcher in Bruges in September 1988 where the Prime Minister stated, "We have not successfully rolled back the frontiers of the State in Britain only to see them re-imposed at a European level."

We now face a more insidious and profound challenge to our liberties – the rising tide of intolerance. The Bruges Group challenges false and damaging orthodoxies that suppress debate and incite enmity. It will continue to direct Britain's role in the world, act as a voice for the Union, and promote our historic liberty, democracy, transparency, and rights. It spearheads the resistance to attacks on free speech and provides a voice for those who value our freedoms and way of life.

The Bruges Group holds regular high–profile public meetings, seminars, debates and conferences. These enable influential speakers to contribute to the European debate. Speakers are selected purely by the contribution they can make to enhance the debate.

For further information about the Bruges Group, to attend our meetings, or join and receive our publications, please see the membership form at the end of this paper. Alternatively, you can visit our website www.brugesgroup.com or contact us at info@brugesgroup.com.

Contact us

For more information about the Bruges Group please contact:
Robert Oulds, Director
The Bruges Group, 246 Linen Hall, 162-168 Regent Street, London W1B 5TB
Tel: +44 (0)20 7287 4414 **Email:** info@brugesgroup.com